D0712899

BACTERIOPHAGES

Bacteriophages

JOHN DOUGLAS

M.Sc. Ph.D. D.I.C. M.I.Biol

Department of Applied Biology

Brunel University

CHAPMAN AND HALL

LONDON

First published 1975
by Chapman and Hall Ltd
11 New Fetter Lane, London EC4P 4EE

© 1975 John Douglas

Typeset by Tek Art
and printed in Great Britain by
T & A Constable Ltd
Edinburgh

ISBN 0 412 12630 3 (cased edition)
ISBN 0 412 12640 0 (Science Paperback)

Distributed in the U.S.A.
by Halsted Press, a Division
of John Wiley & Sons, Inc., New York

Library of Congress Catalog Card Number 75-1288

Preface

The contribution of bacteriophages to the development of modern biology cannot be overestimated yet, sixty years after their discovery, they are as remote and mysterious to many scientists as they are to most laymen. This book endeavours to remedy the situation: an attempt has been made to provide, in readily comprehensible form, a nucleus of information essential to anyone embarking on the study of bacteriophages or using them in their work for the first time. It shows the range of bacteriophage structure and behaviour; it illustrates the role of bacteriophage in molecular biology; it surveys the current state of the art; it presents the medical and industrial aspects. Some simple experimental procedures are given in sufficient detail for the beginner to attempt them successfully. Other, more sophisticated procedures are presented so as to impart a feeling of intimate reality without dazzling the reader with technical complexity.

I hope that young readers will forgive me for assuming that they have some knowledge of bacteria, nucleic acids, antibodies and isotopes. Likewise I would ask mature workers to excuse the omission of cherished specialities. To have included all these, valuable though they are, might have put this book beyond the reach of the phage-novices for whom it is intended. Specific references, save a few of particular interest, have been omitted. Other books on bacteriophage provide them in abundance. They are all much larger and more expensive than this book, which in no way contrives to compete with them. On the contrary, it should stimulate the student to explore them and guide him in his selection. That is not to say that only simple, non-controversial aspects are considered here; that would deny the reader his share of the current excitement of bacteriophage research. Brief summaries of some recent, far-reaching achievements and speculations were considered a necessary part of this irreducible minimum.

I wish to thank all those persons and organizations who have allowed me to use their illustrations; Veronica Phillips for preparing specimens for photography; Fergus Kirkham for some of the line drawings and Dr. Susan E. Smith for assistance with Chapter 7 and reading the final draft.

Brunel University 1974   John Douglas

Contents

1 Introduction

Bacteriophages, usually abbreviated to 'phages', are viruses
that infect bacteria. The first account of a bacteriophage
was published by F.W. Twort in 1915. He had noticed that
some colonies of a bacterium, *Micrococcus*, had undergone
what he termed a 'glassy transformation' in that, instead of
their normal, opaque creamy-white appearance, they had
become clear, like glass. When he transferred a minute speck
of material from such a glassy colony to a normal colony that
too, in a day or so, became glassy. The glassy appearance
could be transmitted from colony to colony indefinitely in
this way and, because Twort could find no bacterial cells in
the glassy colonies and deduced that they must have lysed
(dissolved), he called the phenomenon 'transmissible lysis'.
Material from a glassy colony was found to bring about the
glassy transformation at high dilution, even after it had been
passed through a filter whose pores were fine enough to
restrain the smallest bacteria. Heating the filtrate destroyed
its ability to bring about the transformation. Only colonies of
living bacteria could be lysed by it and during lysis, it was
shown, the lytic agent increased many hundredfold in
quantity. On the basis of this evidence Twort cautiously
suggested that the agent might be, among other possibilities,
a virus similar to those that infect plants and animals, which
had been discovered a few years previously. Little notice was

taken of Twort's work at the time.

The word 'bacteriophage', which means 'bacteria-eater', was coined by F. d'Herelle who seems to have rediscovered the phenomenon independently in 1917. Unlike Twort, d'Herelle had no doubt that it was due to a virus. Furthermore, he soon envisaged the possibility of using phage for the treatment of bacterial infections in Man and animals. This, unfortunately, is not yet possible. Even though bacteriophages are able to annihilate huge populations of pathogenic bacteria in minutes *in vitro,* all attempts to get them to do so *in vivo* have been uniformly unsuccessful. The reasons for this are not clear, even today. It may be that bacterial populations *in vivo* rapidly develop resistance to phage; alternatively the phage, being itself an exogenous substance, may be eliminated by the body's defences before it can significantly affect the pathogen. D'Herelle nevertheless succeeded in arousing world-wide interest in bacteriophage research which revealed a veritable cornucopia of biological discoveries and continues unabated at the present time.

During the 1920's and 1930's many of the fundamental properties of bacteriophages were investigated. Contrary to the belief held by d'Herelle, there was found to be not just one bacteriophage capable of attacking all bacteria but many different phages, each having a limited range of host bacteria that it could infect and lyse. Normally the bacterial hosts of a particular strain of phage are closely related variants of a single species. Any one strain of bacteria is usually susceptible to infection by a group of phages but these may differ widely among themselves in other properties. The physical and chemical factors affecting the growth and death (more correctly 'multiplication' and 'inactivation') of bacteriophages were studied using the plaque assay technique which will be described in full in the following chapter.

Although bacteriophages are too small for their shapes to be resolved in even theoretically perfect light microscopes, it was nevertheless possible to observe phages as dancing points of light in a high-power, dark-field microscope and thus to count the number of particles liberated by the bursting of a single infected bacterial cell. By testing the ability of phages to pass through ultrafine filters of graded pore size, estimates of the dimensions of phages were made which subsequently turned out to be remarkably accurate. Controversy over whether phages were organisms or chemicals in solution, however, continued for many years.

The 1940's saw two developments that gave renewed impetus and direction to bacteriophage biology. Firstly, the development of the electron microscope reached the stage where bacteriophage particles could be seen in clear outline for the first time, confirming their particulate nature beyond dispute and revealing some of their diversity of size and shape. Secondly, whereas previous workers had tended to study in isolation, using a variety of heterogeneous phages, a group in the U.S.A. decided to concentrate on the phages of *Escherichia coli,* in particular those known as the 'T-phages'. This was both a good and a not-so-good thing; good because by standardising their organisms and, to a large extent, their methods communication and cross-fertilization of ideas were facilitated; not so good because the rapid advances in bacteriophage morphology, biochemistry and genetics that gathered pace and continued to the present time relate almost entirely to the phages of *E. coli.* Practically all bacteria that have been investigated have proved to be susceptible to one or more phages and some of these, like the lipid-containing *Pseudomonas* phage PM2 [8], the first shown to have any major structural component other than nucleic acid and protein, have revealed new and hitherto unsuspected properties.

The 1950's could be called the 'Golden Age' of phage biology. Phage studies played a major role in the discovery that deoxyribonucleic acid (DNA) was the primary repository of genetic information, and in the elucidation of the genetic code. Although of minor (but significant) economic import-ance in themselves, phages provided a model of behaviour for researches whereby viruses with greater influence on human affairs could be better understood. Through the 1960's to the present, phages have proved to be ideal material with which to investigate the molecular mechanisms of genetics. If one wishes to breed a better strain of wheat or a faster race-horse, then one must experiment with wheat and horses but these organisms are too complicated for the study of fundamental processes. Over the last seventy years the search for a model system easily susceptible to investigation at the molecular level has led through fruit flies (*Drosophila*), fungi (*Neurospora*) and bacteria (*E. coli*) to bacteriophages, which include the smallest biological entities endowed with the power of self-replication. Phages, indeed, do little other than replicate and therein lies their attraction. Unencumbered by the need to maintain complex life-support systems since these are borrowed from the host cell, the phage genome can comprise up to half the total weight of the particle. Thus there is little to obstruct experimental procedures designed to affect the genome directly and at the same time the very limited phenotype allows isolation of specific aspects for study. Thus genome and phenotype can be related in a way that is not possible with cells. The bacterial hosts are very conveniently maintained *in vitro* under controlled conditions requiring relatively little in terms of time, space and labour, in contrast to the demands of plants and animals. Thus within the space of a day one phage particle may be multiplied into billions; these can be treated with a mutagenic chemical and the rarest of mutants detected by examining the entire population — all on a few square feet of bench space. The ability to handle huge populations economically is a consequence of bacteriophages' small size yet, if required for biochemical analysis or manipulation, modern fermenter technology permits gram quantities of bacteriophages or their isolated nucleic acids, proteins and enzymes to be prepared.

For most purposes, until comparatively recent years, the phage phenotype consisted of the macroscopic appearance of the phage-infected bacterial culture. However, other biological entities may produce similar appearances and it is important for the phage student to be aware of these. Both *Bdellovibrio bacteriovorus* and colicin K can produce local clearings in layers of otherwise confluent bacterial growth and both can cause the turbidity of broth cultures of bacteria to be reduced. But there the resemblance ends. *Bdellovibrio* is a cell, able to move, respire, increase in size and reproduce by fission. Colicin K is simply a protein molecule unable to reproduce itself at all. Bacteriophages, as Figure 1.1 shows, are intermediate in size between these two. They are also intermediate in complexity. Are they to be regarded then as the present-day representatives of an intermediate stage in evolution? It is true to say that, at the lower end of the

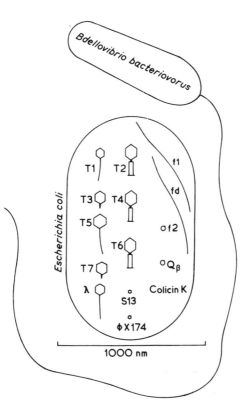

Figure 1.1 Relative Sizes of a Bacterium (*E. Coli*) and an Assortment of Biological Entities (*Bdellovibrio*, Bacteriophages and a Colicin) which attack it.
Colicin K is beyond the resolution of current electron microscopes. The size depicted is based on indirect estimates.

biological size range, smallness necessitates simplicity of structure and tends to be thought of as primitive. But was the small size of phages arrived at by the elaboration of an even smaller molecule, did they arise by the simplification of a cell down to the barest requirements for a parasitic life, or did they travel some evolutionary route that is exclusively their own? Some clues may be gleaned from the following pages but until the extent of the bacteriophage world is more fully explored it would be best to reserve judgement on this question.

2 Lysis

Lysis, that is the dissolution of the bacterial cell, is the essential phenomenon whereby the presence of phage is detected. Some filamentous phages simply leak out of the cell over a period without killing it; certain defective phages multiply in the infected cell but cannot lyse it; some phages termed 'temperate' may integrate themselves into the bacterial cell and lyse it only when induced to do so by special treatments. In such cases phage is detected only with difficulty and if all bacteriophages behaved in any one or other of these ways they might not yet have been discovered. Fortunately the virulent, lytic phages are very common. They lyse the cell in which they have multiplied by causing the production within it of an enzyme, lysozyme, that attacks the murein of the cell wall, weakening it so that it bursts and liberates the phage within. In some cases the dissolution of the wall is very nearly complete, in others a substantial amount of debris remains. Lysis can be detected in several ways.

2.1 Lysis on solid media

A plate of nutrient agar can be inoculated heavily and uniformly with host bacteria by flooding its surface with a broth culture, removing the excess liquid and allowing the plate to dry. When incubated, a 'lawn' of bacterial growth will develop. If drops of concentrated phage suspension are placed on the plate soon after the inoculum has dried, the lawn that develops will have bare patches corresponding to where the drops lay. If instead a few phage particles are mixed with the inoculum before it is applied, *each particle* will produce a hole in the lawn in the following way: it first of all infects a single cell, multiplies within it and causes it to lyse, liberating many progeny phages; these diffuse away from the original site, infect other cells and lyse them and the process is repeated so that the area of lysis, known as a 'plaque',* grows to be visible eventually to the naked eye. These simple techniques have their uses but there are problems in ensuring the even distribution of inocula and the time they take to dry. More reproducible results can be obtained if the host bacteria and the phage are embedded in a thin layer of agar. Phages can diffuse through agar gels to form plaques. If a low concentration, for example 0·4 per cent, of agar-agar is used the gel offers less resistance to diffusion and plaques are larger. Such agar, known as 'semi-solid', 'soft' or 'sloppy' agar, is normally employed as a thin layer (c. 1 – 2 mm) poured over a thicker layer of normal solid agar which provides, by diffusion, the main source of nutrients for the bacteria in the sloppy layer.

By counting the plaques that arise from a known volume of phage suspension one can estimate the number of phage particles in it. This is known as the 'plaque assay technique' and the full procedure is shown diagrammatically in Figure 2.1. It is simple, rapid and accurate. It is probable that without the plaque assay phage research would hardly have advanced beyond Twort and d'Herelle's original observations. A similar technique for animal viruses was not developed until 1952 and one for plant viruses is still lacking, which goes a long way towards explaining why these branches of virology lag so far behind phage.

Plaque formation by phages is not 100 per cent efficient; that is to say, not every particle added to a lawn of bacteria

---

* In some virologists' vocabulary this word can also be used as a transitive verb meaning "to cause to make plaques". Others use the verb "to plate" synonymously. Unfortunately microbiologists in general use "to plate" meaning simply "to inoculate (bacteria or fungi) into a dish of agar".

forms a plaque. Some may infect dead bacteria for example, in which case they are unable to develop further; others encounter random hazards during manipulations that damage them. Under favourable circumstances some phages may approach 100 per cent plaquing efficiency but others have an intrinsically low efficiency of 10 per cent or less because of innate defects.

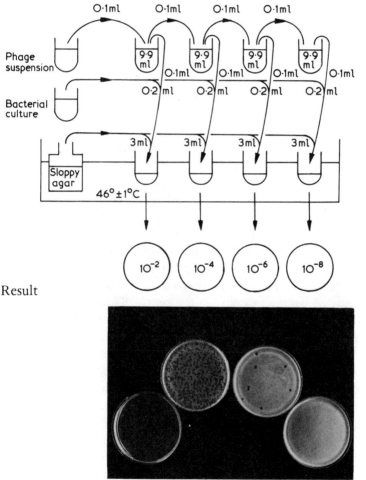

Result

Figure 2.1 Phage Assay by the 'Sloppy Layer' Plaque Technique
The phage suspension was serially diluted by transferring 0.1 ml portions through 9.9 ml portions of diluent, making 10^{-2}, 10^{-4}, 10^{-6} and 10^{-8} dilutions of the original. Four sterile, empty test-tubes

were placed in the waterbath at $46°C \pm 1°C$ and 0.1 ml of the
corresponding phage dilution added to each. To each was then added
0.2 ml of an overnight broth culture of the host bacterium and 3 ml of
'sloppy agar' (broth + 0.4 per cent agar-agar), which had been melted
and cooled to $46°C$ in the waterbath. The contents of each tube
were then well mixed by swirling, poured over the surface of a plate
of normal, solid agar and allowed to set level. The plaques were
visible after six hours. The titre of the original suspension, calculated
from the 10^{-6} dilution, is $8 \times 10 \times 10^6 = 8 \times 10^7$ *plaque-forming
units* (p.f.u.) ml^{-1}.

Plaques have other uses besides assay. The plaque characteristics are often useful in distinguishing different phages. The
size may vary within limits for any particular phage but
whereas some phages typically produce plaques of 5 mm
diameter or more, others rarely exceed 1 mm. The shape of
most plaques is circular but mutants producing sectored or
'star' plaques have been described. The plaque margin may
be sharp or diffuse and there may be a zone of turbidity
surrounding it (Figure 2.2). If a phage is plaqued using a
mixture of two different strains of bacteria the plaques will
be clear only if both strains are lysed. If only one of them
is lysed the plaques will be turbid due to the growth of the
other. This 'mixed indicator' technique is useful for isolating
host-range mutants. Sometimes a plaque is seen to have a
colony of bacteria exactly in the centre. This may be a
'lysogenic' colony that secretes phage to which it is immune
but the rest of the lawn is not.

The phages in a plaque usually constitute a clone, a population of genetically uniform individuals that have arisen
from a single individual by vegetative reproduction. Thus a
pure culture may be obtained from a mixture of phages by
plaquing it, taking a few particles from a well-isolated plaque
by stabbing it with a wire, and using them to initiate a new
culture.

An assay plate showing almost confluent plaques may be
used to prepare a concentrated phage suspension. The sloppy
layer is scraped off, macerated with water or other diluent
and the agar and bacterial debris removed by low speed
centrifugation. Suspensions with a titre of over 10^{12} plaque-

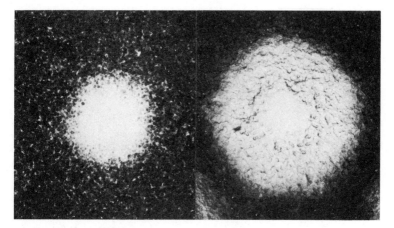

Figure 2.2 Diffusible Lysin
The plaque on the left (*Staphylococcus aureus* phage 42D) has a
well defined edge. That on the right (*Streptococcus lactis* 0712) is
surrounded by a halo of partially lysed cells, caused by the soluble,
lytic enzyme that diffuses from the centre faster than the phage particles
themselves Both plaques were grown in sloppy agar by the double-
layer technique and photographed at a magnification of 40x after
24 hours. The background lawn can be seen in each case to be
composed of crowded *colonies* of bacteria; the individual cells
cannot be seen at this magnification.

forming units (p.f.u.) per ml can often be prepared by this
method. It is not always necessary to remove the sloppy
layer. Adequate titres may sometimes be obtained simply by
washing its surface with a few ml of diluent.

Phage plaques are often discernible, albeit with some
difficulty, after as little as five hours incubation. Surprisingly,
they may not increase in size after that time even though
they become much clearer. A few combinations of phage,
host and medium result in plaques that appear to grow
indefinitely but these are rare. No satisfactory general
explanation as to why plaques stop growing has been advanced.
Depletion of nutrients for bacterial energy requirements and
biosynthesis and the accumulation of toxic metabolites may
explain some, but not all cases. Microscopy of the plaque
margin is interesting. The light microscope often reveals

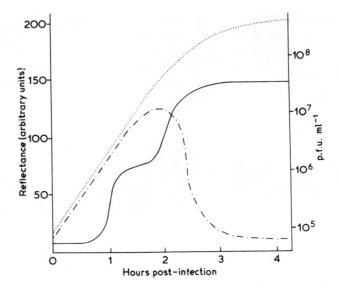

Figure 2.3 Lysis in a Broth Culture
This shows what may be expected when sufficient phage is added to
a growing culture to infect about 1 per cent of the cells. The growth
of an uninfected (control) culture (dotted line) levels out as the
medium becomes exhausted. The infected culture (pecked line)
appears to begin similarly since only about 1 per cent of the cells
are affected; this level of infection will hardly register at all on a
nephelometer (Fig 2.5 B). After about an hour, however, these cells
burst and liberate enough phage to infect the other 99 per cent.
Growth then stops and eventually the whole culture clears. The
two cycles of infection are revealed more clearly if the number of
plaque-forming units (p.f.u., = phage particles plus infected cells) is
simultaneously followed (solid line). The second 'step' is less sharp than
the first since the progeny of the first cycle of infection are not all
released at the same time and thus the second cycle is not synchronous
in all cells. No third cycle is possible since there are no host cells left.

Although this simple experiment is instructive for the beginner, it
yields little precise information of value to the phage biologist.
Firstly the presence of uninfected cells at the end of the first cycle
means that there is a loss of p.f.u. due to two or more phages being
adsorbed onto the same cell; thus one does not know how many
were released. Secondly the continued growth of the uninfected
cells, superimposed on the growth of the phage, confuses the
biochemical picture. The more sophisticated 'one-step growth'
experiments, in which a single cycle of replication is isolated for
study, are shown in the following figure (2.4).

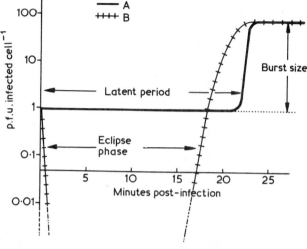

Figure 2.4 One-step Growth Experiments

A.   Normal Lysis (heavy line). A broth culture of the host
bacterium, growing in the logarithmic phase, is inoculated with
phage at an input ratio of one phage per 10 cells to ensure that
each cell that is infected is infected with only a single phage since
multiple infection may give a different result. After 5 minutes the
mixture is diluted in broth containing specific antiphage serum. The
serum inactivates any phage still free; dilution itself reduces the
probability of further adsorption so that one is dealing thereafter
with a culture in which infection is nearly synchronous. Five minutes
later the culture is diluted again to inactivate the antiserum. Samples
are removed for plaque assay every few minutes.

Initially each phage gives *one* plaque. When the phage adsorbs
onto and infects a cell, that too will give *one* plaque. The number
of plaque-forming units per cell remains unity during what is called
the 'latent period' up to the onset of lysis. When a cell lyses it
releases a large number of phages, each of which can produce a
plaque, so that the number of p.f.u. per cell shows a massive
stepwise increase. The relative increase, representing the average
number of phages released from one cell, is known as the 'burst
size'. Because of the great dilution of the original mixture, the
probability of loss of p.f.u. by multiple adsorption is very small.

B.   Premature, artificially-induced lysis (barred line). This experiment
is done to find out what is going on inside the infected cell during
the latent period. The technique is similar to that in A. but when
samples are taken for assay they are first treated so as to break open

the cells and liberate any phages they contain. Of the many methods
of breaking open cells without inactivating the phages they contain,
shaking with chloroform has been most widely used. It is very
effective with *E. coli* and similar enteric bacteria but with others,
staphylococci for instance, it is totally ineffective and other methods,
such as mechanical rupture or explosive decompression, must be
used. Within a few minutes of infection the phage virtually disappears
from the culture; one cannot extract p.f.u. from the cells by any
method. This state of affairs persists until shortly before normal
lysis would occur. The p.f.u. increase rapidly in number until they
reach the level that normal lysis would produce. The period during
which no phage is extractable is known as the 'eclipse phase'.

grossly swollen and distorted cells. Electron microscopy may
reveal very few phages at the margin, suggesting that the
latter stages of plaque growth are due, not to phage but to
the diffusion of excess lytic enzyme from the central area.
A few unlysed cells are usually to be seen scattered through-
out the plaque.

2.2 Lysis in liquid media

This can be the most dramatic manifestation of phage activity.
If a growing broth culture is inoculated with a few particles of
a phage to which it is susceptible ("homologous" phage) the
culture appears to continue growing normally for some hours
(or minutes, depending on the 'input ratio' of phages to
bacteria) then, in the space of as little as perhaps ten minutes,
the whole culture becomes glass-clear as the cells lyse. What
one is witnessing is the progressive multiplication of the
phage, imperceptible at first since only a few cells are affected
but, as they lyse and liberate progeny phage, the proportion of
infected cells rises exponentially until the last cycle of lysis
involves all that remain (Figure 2.3) so that clearing is sudden
and complete. Figure 2.5 shows instruments used to quantify
lysis in liquid media more precisely than purely visual
observation allows. The *absorptiometer* measures turbidity
as the impedance offered to light passing *through* the suspension of organisms; the *nephelometer* measures turbidity as the

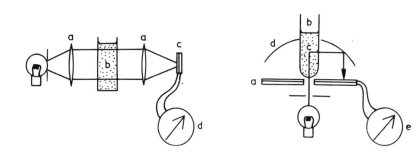

Figure 2.5 Instruments for Measuring the Turbidity of Bacterial Suspensions

A   Absorptiometer. A beam of light, collimated by lenses a, a, passes through the suspension in cuvette b and impinges on photo-cell c, thereby generating an electromotive force registered on galvanometer d. In modern instruments the galvanometer scale is usually calibrated in units of extinction relative to a control solution. The relationship between cell concentration and extinction obeys the Beer-Lambert law fairly closely over a wide concentration range.

B   Nephelometer. The path of a single ray is shown. It passes through the aperture of annular photocell a into culture tube b and strikes a bacterium at c which deflects it through $90°$. It is then reflected by hemispherical mirror d onto the surface of the photocell and generates an electromotive force which is registered on galvanometer e. The galvanometer scale is calibrated in arbitrary units relative to a stable standard (usually a ground glass rod) and the reading is known as the 'reflectance'. The relationship between cell concentration and reflectance is complex.

   The optical properties of bacterial suspensions vary with the species and physiological condition. Both nephelometers and absorptiometers require a new calibration curve for each set of circumstances. Rectilinear relationships are not always obtained, particularly at high bacterial concentrations where forward scattering and reabsorption become significant. Curvilinear relationships may be perfectly adequate for practical purposes. In suspensions that are undergoing lysis and contain swollen and ruptured cells, the transla-tion of optical data into cell numbers is of dubious value; these instruments are nevertheless very useful for determining the times of onset and completion of lysis and *relative* rates of change.

amount of light deflected *sideways* by the cells. The choice of
instrument depends on many factors. One advantage of
nephelometers is that they are designed, as a rule, to accept
the culture in its normal tube instead of requiring it to be
transferred to a special cuvette as in most absorptiometers.

2.3 Lysis of individual cells

Lysis of individual cells by phage may be observed with the
light microscope provided steps are taken to immobilise them.
This is conveniently done in the following way. A culture is
infected with phage and the turbidity followed with one of
the instruments mentioned above. When the turbidity ceases
to rise, indicating that massive lysis is imminent, a small
loopful (*c.* 0.001 ml) is removed and placed on a thin layer
of agar on a slide. The liquid soaks into the agar in seconds,
leaving the infected bacteria perched immobile on the surface.

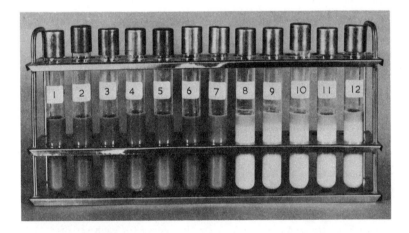

Figure 2.6 Assay by 'Dilution to Extinction'
In the example shown, 1.1 ml of a suspension of *Streptococcus
lactis* phage 0712 was added to the first 9.9 ml portion of litmus
milk in tube No. 1. 1.1 ml of this mixture was then transferred to
tube No. 2 and so on, making serial decimal dilutions down to
10^{-12}. One drop of a $\frac{1}{100}$ dilution of an overnight broth culture of
Streptococcus lactis strain 712 was then added to each tube and they were
incubated at 30°C overnight. Next day tubes 1 − 7 still showed the

characteristic purple colour of uninoculated litmus milk but tubes
8 – 12 showed the pink acidic surface and bleached lower part
typical of the growth of *Streptococcus lactis*. Tube No. 7 contained
1 ml* of the previous, 10^{-6}, dilution. As it is the last tube in the
series to show inhibition of bacterial growth, that 1 ml must have
contained at least one phage particle. The titre of the original phage
suspension was therefore in the order of 10^6 particles per ml. This
technique is simple and instructive for beginners but it is
arithmetically 'messy', statistically naïve and lacking in precision. With
even 50 replicates at each dilution its accuracy would still be only about
that of a single plaque assay. It is useful for making rough comparisons
of the amounts of phage in different preparations under circumstances
where plaque assay is not practicable but its main use is not as an assay
but for propagating phage from minimal inocula, by taking the last
tube before extinction to make a stock. The National Collection of
Dairy Organisms recommends it for this purpose. In the example
shown, the litmus milk in tube No. 7 would have to be decaseinated by
acidifying it with dilute lactic acid, centrifuging to remove the casein
thereby precipitated and carefully neutralizing the supernate for stock.
The use of litmus milk is usual for dairy phages. Dilution to extinction
can, of course, be used for any phage having an overt effect on the
host culture.

* It received 1.1. ml but 0.11 ml was removed in making the next dilution so
the volume remaining was actually 0.99 ml. So long as one is using the same
portion of liquid for making *and* testing a dilution, it is not possible to choose
transfer and diluent volumes that will make both the dilution-ratio and the
residual volume convenient round figures. By calling 0.99 ml 1 ml, an error of
1 per cent is introduced. Considering the low intrinsic accuracy of this assay
and the uses to which it is put, this error is not significant.

They can be covered with a coverslip and examined continuously until they lyse, using either a dark-field or phase-
contrast microscope. A high magnification, 1000 x at least,
is required if details of the rupture of the cells are to be
perceived but if all one wishes to see is the sudden disappearance of each cell 400 x is quite adequate. Cells in the terminal
stages of infection are little affected by small changes in
temperature and the time at which the broth culture in a
waterbath at, say, 30°C lyses is found to correspond well
with the disappearance of cells in the sample under the
microscope.

2.4 Stock lysates

A lysate may serve as a stock of phage for use in subsequent experiments. The highest possible titre is desirable. This is usually obtained by the sloppy agar method described above but varies somewhat from phage to phage. Lysis of a stationary broth culture can give yields of 10^9 p.f.u. ml^{-1} or more and this may be increased by a factor of $10-1000$ if the culture is aerated in a bubbler tube. A typical bubbler tube is shown in Figure 2.7. It is normally used for volumes up to 25 ml but there is no limit to potential escalation.

The lysate, however prepared, is centrifuged at low speed or passed through a bacteriological filter or both, to remove unlysed bacteria and debris. Filters of the 'Millipore' membrane type are preferable to thick pads since they retain less phage. Sometimes chloroform is added as a preservative but storage at low temperature ($+4°$C) is usually adequate. Freezing is not normally necessary and may even result in loss of titre. There is much to be said for dispensing the stock phage into small ampoules containing about 1 ml or such other quantity as will suffice for one experiment. On this 'use and discard' system the possibility of the growth of contaminants in an opened stock bottle between experiments is eliminated.

2.5 Bdellovibrio (Figure 1.1)

Bdellovibrio is not a phage but it is included here because it is bacteriolytic and can produce plaques in a lawn of bacteria like a phage. It is a very small bacterium, just recognisable as such in the light microscope, which parasitises other bacteria, usually gram-negative rods, and causes them to lyse. It can be plaque assayed like a phage but since plaques of *Bdellovibrio* take $3-5$ days to appear there should never be any confusion of the two. The name is derived from Latin words meaning 'leech' and 'shaker', which describe its modes of life and locomotion very well.

It swims very rapidly with its single polar flagellum; the initial contact with the host is violent and the host cell is said to recoil. The parasite attaches itself by its front end and

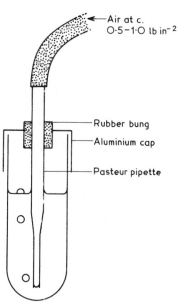

Figure 2.7 Bubbler Tube
Bubbler tubes are widely used for propagation of phages of aerobic
bacteria. A high rate of aeration ensures maximum available energy for
growth and reduces the accumulation of toxic metabolites. Several
tubes of the pattern shown may be worked simultaneously off a small
aquarium aerator. A cotton-wool filter may be inserted in the airline or
in the pasteur pipette to protect the culture from contamination.
Exhaust air escapes under the rim of the cap; it will contain phage and
bacteria in an aerosol; if this is undesirable (when a pathogen is used,
for instance) the pipette may be introduced through a firm cotton-wool
plug but convenience is lost.

squeezes into the host cell through a hole smaller than its
own diameter which it makes, presumably, by enzymic means.
Once inside, *Bdellovibrio* appears to draw nourishment from
the cytoplasm without actually penetrating it since, in
electron micrographs of sections of infected bacteria, it can
be seen between the cell wall and the membrane. Cell division
occurs within the host, which eventually lyses. *Bdellovibrio*

can be maintained in broth cultures of host bacteria and separated from them by means of a 'Millipore' filter of 0.44 micron size, through which the parasite can pass but normal bacteria can not. *Bdellovibrio bacteriovorus* and host strains of *E. coli* are maintained in the National Collection of Industrial Bacteria (see Appendix C).

3 Structure and function of the virion

The infectious phage particle, like the particles of other viruses, is called a 'virion'. Figure 1.1 shows in outline the virions of a range of phages of *E. coli* ('coliphages'), illustrating the diversity in size and shape of the virions of phages of a single bacterial species. On analysis, practically all phage virions are found to consist of protein and nucleic acid only. The nucleic acid is commonly said to be exclusively either DNA or RNA but this may not be strictly true. One coliphage, T5, contains mainly DNA but probably some RNA-like material as well [13].

The elucidation of phage structure and its correlation with function is a slow process. Coliphage T2 has probably been investigated in more detail than any other but is not to be regarded as typical in any way. It will nevertheless be advantageous to study it first and use its properties as reference points enabling the virions of other phages to be seen in perspective.

3.1 Coliphage T2

This virion consists of a polyhedral head approximately 100 nm x 80 nm and a tail about 100 nm long joined to it by a short 'collar piece'. At the distal end of the tail is a roughly hexagonal 'end plate' bearing at each corner a short 'tail pin' and a long, jointed fibre. The main part of the tail is composed of two concentric tubes known as the 'core' and 'sheath' respectively. The head comprises an outer protein 'membrane' enclosing a central mass of tightly packed DNA. The manner of packing of the DNA is still in dispute. X-ray diffraction data suggests that it is arranged in parallel bundles; some recent electron micrographs [12] of partially disrupted phage heads indicate that it is wound in a ball, as if on a spindle. The two viewpoints are not mutually exclusive. What is beyond dispute is that packing must be orderly, not haphazard. All the evidence, microscopical, biochemical and genetic, shows quite clearly that T2 phage DNA has the form of a single strand, an unbranching linear molecule more than 500 times as long as the head that contains it, and that during infection the DNA passes along the narrow central channel of the tail core. A very special packing arrangement is surely necessary for this to occur smoothly, without knotting or tangling.

There is evidence that some phage DNA's, under certain conditions, are circular. The word 'circular' is not used in the strict geometric sense but implies that the ends of the linear molecule are joined so that it forms a closed loop, which is endless, like a circle. Phage *lambda* (λ) is one such phage. In the virion its DNA is linear but on entering the host cell the ends join to form a circle (Figure 5.2). The DNA of phage ØX174, however, is circular both in the virion and in the host cell. It has been suggested that the DNA of T4 (a close relative of T2) must circularise in the host cell in order to replicate. Evidence for this view at the moment is tenuous.

How does the T2 virion find and attack a cell? In aqueous medium the virion is constantly bombarded by moving water molecules that keep it in a state of continuous rapid motion. The tail fibres can be envisaged as flailing in all directions under the same influence. When it chances to come within range of a suitable host, one of the phages tail fibres will stick to it. As the phage and bacterium are then kept in close proximity by this attachment it is a matter of moments only before all the fibres have adhered. The phage is then said to have 'adsorbed' to the bacterial cell, reversibly at first but irreversibly soon afterwards. The end plate of the virion attaches to the bacterial cell wall and the tail sheath contracts so that the tail core penetrates the wall which may be weakened by enzymes contained in the core canal. The nucleic acid of the phage is then injected into the lumen of the cell through the hollow tail.

Figure 3.1 Coliphage T2
Negatively stained. (Courtesy of Dr. M. Perutz, Medical Research Council Laboratory of Molecular Biology)

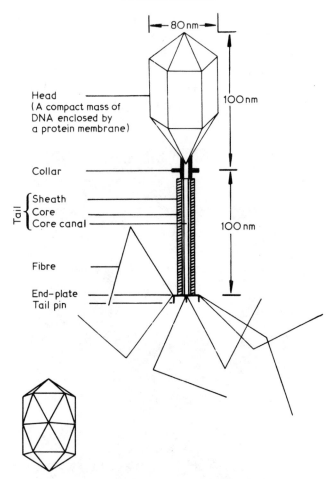

Figure 3.2 Anatomy of the Coliphage T2 Virion

The principal parts are shown in their normal relation to one another.
The head is depicted so as to emphasise its polyhedral nature; the
geometry is based on its appearance in electron micrographs but
theoretical considerations suggest that the true shape is more complex
(inset). The tail is shown in section revealing the helically constructed
sheath around a hollow core. Dimensions are approximate. Older
estimates based on filtration and sedimentation are mainly of historical
interest now but dimensions in electron micrographs also vary. This
could be due to strain differences but is more likely due to the techniques of preparation. During drying the head flattens and spreads to a
varying degree. The proteinaceous components collectively constitute
the 'capsid' of the virion.

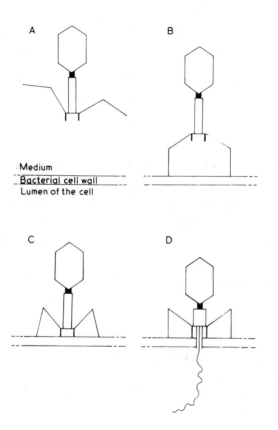

Medium
Bacterial cell wall
Lumen of the cell

Figure 3.3 The Mechanics of Infection by Coliphage T2
A.   Free phage. B. Adsorption by tail fibres. C. Tail pins engage cell
wall. D. Sheath contracts, core penetrates cell wall and nucleic acid
molecule is injected.

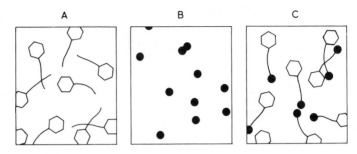

Figure 3.4 Adsorption of Coliphage T5 to Purified Receptor *In Vitro*
A. Coliphage T5. B. Receptor molecules isolated from the cell wall of *E. coli* B. C. Mixture of A and B after incubation. (based on electron micrographs by W. Weidel *et al*)

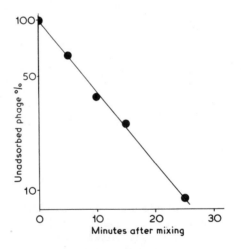

Figure 3.5 The Kinetics of Adsorption
A suspension of *Streptococcus lactis* 712 containing about 10^8 cells per ml was made in broth and about 10^7 p.f.u. of phage 0712 per ml was added. 1 ml portions were removed at once and at intervals thereafter, diluted 1:100 in chilled broth (to prevent further adsorption and extend the latent period of the phage), the bacteria and adsorbed phage removed by centrifugation and the unadsorbed phage in the supernatant determined by plaque assay. (B. Cartwright, *unpublished data*).

Adsorption is very host specific and depends on the presence in the cell wall of specific receptor sites. In the case of another phage, T5, the receptor has been isolated. It is a lipoprotein molecule of high molecular weight. When phage T5 and its receptor are incubated *in vitro*, electron microscopy shows that the receptor molecules become attached to the tip of the tail. In phage T4 the tail fibres are wound tightly around the tail and cannot function in tryptophan-free media. Tryptophan allows the fibres to unwind.

Many phages have a very definite requirement for divalent cations particularly calcium, which promotes adsorption and without which infection may fail to occur at all. This explains why they are inhibited by, for example, citrates and oxalates which form insoluble compounds with calcium in the medium and render it unavailable. Ethylene-diamine-tetra-acetic acid (synonyms: EDTA, 'Versene') chelates calcium and is highly inhibitory to calcium-requiring phages. Magnesium may be used as a replacement for calcium by some but not all phages with this requirement.

The contraction of the T2 tail sheath, driving the core through the cell wall, is a most remarkable process that has been intensively studied. It is very doubtful whether mechanical force alone is sufficient to effect penetration. If a large number, say 100 virions, are allowed to adsorb simultaneously onto a single cell it lyses immediately. This is known as 'lysis from without' (phages normally escape from the host cell by 'lysis from within' but the term is hardly ever used). It suggests the presence of an enzyme similar to lysozyme in the tip of the phage tail to assist penetration. The 'trigger' mechanism that controls sheath contraction can be set off artificially by treating the phage with either hydrogen peroxide or zinc cyanide. The natural stimulus is unknown. The sheath of T2 is composed of 144 comma-shaped protein subunits arranged in a helix having twelve turns, each of twelve subunits. On contraction a rearrangement occurs giving six turns of 24 subunits, thus reducing it to half its original length. This is an energy requiring process. The tail contains 144 molecules of ATP which are converted to ADP during contraction. It also contains calcium and folic acid, whose function is uncertain. The passage of the nucleic acid contained in the phage head through the hollow tail core into the bacterial cell deserves detailed consideration.

Table 3.1 The essential strategy of Hershey and Chase's (1952) experiments.

	Expt. 1	*Expt. 2*
Bacteriophage propagated in medium containing radiolabelled . . .	Phosphorus (P^{32})	Sulphur (S^{35})
Radiolable appears in the phage's . . .	Nucleic acid	Protein
When such a phage is used to infect bacteria, then sheared off and the cells separated from the medium by centrifugation, most of the label is found in the . . .	Bacterial cells	Supernatant
Hence, at the moment of shearing, the labelled material was . . .	Inside the cells	Adhering to the exterior of the cells.

It was established by Hershey and Chase in 1952, in a series of experiments that has been widely and, according to one author [16], frequently wrongly cited, that shortly after infection it is possible to remove most of the phage protein but hardly any of the phage nucleic acid from the infected bacterium by vigorous mechanical agitation in a 'Waring Blendor', an instrument resembling a domestic liquidiser. The subsequent multiplication of the phage is unaffected. Clearly, the protein can be removed since it comprises the head sheath and tail which remains attached to the outside of the cell. The nucleic acid cannot be

Figure 3.6 Adsorbed Phage
Part of a section of *Bacillus subtilis* showing three phage SP50θ virions adsorbed to the wall. The head of the central phage appears dark since it still contains nucleic acid; that on the right has lost its nucleic acid and appears 'ghostly'. (Courtesy of Dr. R.G. Milne and Plant Pathology Laboratory, Rothamsted)

removed by agitation ('sheared off') because it has entered the cell. This interpretation is supported by electron microscopy. Phage heads are usually dark in electron micrographs because their tightly packed nucleic acid is not easily penetrated by electrons. Phages whose heads have lost their nucleic acid appear light in electron micrographs and are known as 'ghost phages'. Adsorbed phages frequently appear

ghostly. If a phage suspension is suddenly diluted in hypo-
tonic medium the resultant osmotic shock liberates the nucleic
acid into the medium. However it is very doubtful whether
any such effect could propel the nucleic acid through the
tail and into the host cell; in osmotic shock the nucleic acid
is presumed to be lost through cracks in the head. Neither
is contraction of the phage head sheath a possible propulsive
mechanism since the diameters of free, intact phages and
adsorbed ghosts are not significantly different. The most
satisfying explanation of the movement of the nucleic acid
is provided by the following thermodynamic argument.
If a small length of nucleic acid protrudes from the phage
tail into a large space such as the lumen of a bacterial cell, it
is free to move under the influence of bombardment by
rapidly moving water molecules. The only restraint on its
movement is the fact that it is still attached to the rest of it
contained within the phage. Thus the end can acquire
momentum and, as it reaches the limit of its travel, that
momentum will be expressed as a tug on the rest, pulling
more of it through the tail. Were the phage head and the
bacterium the same size, similar movements inside the
phage head would pull it back and an equilibrium might be
established. The head being so much smaller, however, the
nucleic acid is restrained by its own dense packing which
ensures that it cannot acquire much momentum and thus
quickly enters the larger space. This explanation is satisfying
because it is complete in itself and no special propulsive
organelle or chemical reaction need be postulated; transfer is
the inevitable consequence of the observed structures. There
is, nevertheless, the problem of explaining how the first bit
gets through to pull the rest. Leaving aside the matter of the
nature of the force that propels the nucleic acid molecule,
a model is needed to show how the strand of nucleic acid
may be packed tightly in the head yet in such a way that the
end can find the small opening into the hollow tail and the
rest of this extraordinarily long molecule can follow it in a
minute or so, as it is known to do. On the first point, it may
be that the end of the nucleic acid molecule is already in
the lumen of the tail by virtue of the tail having been

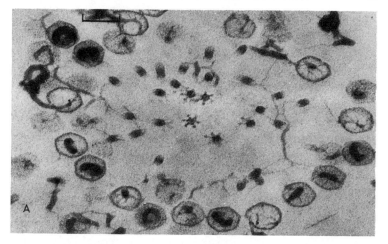

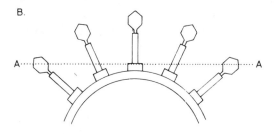

Figure 3.7 Phage End-plate
A. Tangential (grazing) section of *Bacillus subtilis* infected with numerous SP50θ phage particles. The section shows heads, tails and, in the centre, the star-shaped 'foot' or end-plate. Scale-bar = 100 nm. (Courtesy of Dr. R.G. Milne and Plant Pathology Laboratory, Rothamsted)
B. Drawing to show the plane of the section A – A.

assembled round it when the virion was formed. Regarding the second point, one is tempted to think of those commercially available balls of string, where one can pull string endlessly from the centre without rotating the ball, and speculate on similarities.

In even the best electron micrographs the head sheath or "membrane" of T2 appears to be smooth and structureless.

In the light of modern knowledge of nucleic acid function
this is mysterious. The ratio between the weight of a finite
quantity of nucleic acid and that of the molecule of protein
for which it can code is, very roughly, 9 : 1. If the protein of
a virion were all one molecule with a unique amino acid
sequence, then 90 per cent of the weight of the virion would
need to be nucleic acid. Yet even in those virions that contain
the greatest proportion of nucleic acid it never accounts for
as much as 50 per cent. The paradox is resolved if the protein
of the virion is present, not as a single molecule but as many
identical smaller molecules (subunits). This was realised soon
after the theory (now an accepted fact) of the genetic function
of nucleic acid was proposed and the finding that, in the very
smallest phages, a protein subunit structure could indeed be
revealed by electron microscopy provided valuable support
for the theory. Using the technique of negative staining, not
only could the knobs or 'capsomeres' that compose the shell
or 'capsid' of phages such as ØX174 and f2 be clearly seen
but also the 'monomers' from which the capsomeres themselves
are composed. A subunit structure, you will remember, is
known to be present in the tail of T2 — it can be seen with
the electron microscope — so the homogeneous head presents
a problem for which two possible answers have been debated.
One is that it is composed of subunits that are beyond the
limits of resolution of the electron microscope. It is difficult
to see how the precise geometry of the head could be maintained by the large number of subunits required unless the
bonds between them were unusually rigid. The alternative
is to assume that large subunits are present, maintaining the
shape of the head but invisible because they are 'plastered
over' by a layer of different, much smaller protein molecules.
Analysis of the head sheath proteins is not easy and is complicated by the existence of a small amount of internal
protein, possibly involved in the organization of the packing
of the nucleic acid. In the heads of certain highly defective
mutant phages ('polyheads') a subunit structure can indeed
be seen.

3.2 Other tailed phages

The contractile tails of coliphages T2, T4 and T6 (the 'T-even' series) are virtually identical and there are many other phages of other bacteria whose tails are similar although they may vary in length and details of the collar, end-plate and fibres (if any). The contractile tail is a very efficient mechanism and those phages that have it tend to be adsorbed rapidly.

There are, however, several other types of tail. Coliphages T1, T5 and *lambda,* for example, have thin, flexible tails with no sheath which are non-contractile. Coliphages T3 and T7 have just short, tapering stumps. Tails like these are quite common and while there is no doubt that they serve as organs of adsorption and injection it must be admitted that we have little knowledge of how they function.

3.3 Tail-less ('Minute') phages

All known tail-less phages are parasites of *E. coli* and other Gram-negative rod-shaped bacteria. They are spherical, with diameters in the range 22 nm − 25 nm. They are the smallest forms of life and show the ultimate adaptation to smallness in that their genetic material (DNA or RNA) is single-stranded. The nucleic acid must code for the protein that surrounds and protects it but, with diminishing size, the ratio of surface to volume increases and the coding problem becomes more acute. The number of structural proteins in these minute virions may be just two. One of these may function also as a lysozyme but, in addition to these, the nucleic acid must code for the enzyme that catalyses its replication in the host cell. The DNA in higher organisms, as well as that in the larger phages, consists of the familiar double helix. One of the two strands in such molecules is clearly redundant so far as information storage is concerned since, given either strand, the other could be formed upon it. By eliminating the complementary strand the minute phages encode their genomes in the most efficient way possible.

In phages ØX174 and S13 the genetic material is single-stranded DNA. This excited great interest when first discovered but has since been found in several other viruses as well. Single-stranded DNA differs from ordinary double-stranded DNA in several important properties. Firstly the radiation sensitivity is much greater since, when a quantum of radiation causes the single strand to break, there is no complementary strand to hold the broken ends in juxta-position and so increase the chance of reunion. Secondly the double helix has rigidity and tends to be brittle and break up when exposed to the high shearing forces encountered when a solution is pipetted through a narrow orifice; single-stranded DNA is flexible and resists pipetting damage. Perhaps the best evidence for single-strandedness was the finding that in such DNA the ratios of the bases adenine: thymine and guanine: cytosine are not unity, as they should be if paired in complementary strands. Such evidence is hard to obtain since it requires that a substantial quantity of pure nucleic acid be prepared for analysis. Whereas specific inhibitors of RNA and DNA synthesis may be used to differentiate RNA and DNA viruses, there are no such re-agents to distinguish single-stranded from double-stranded DNA viruses as yet. There are, however, fluorescent staining tests [2] that enable nucleic acid type to be determined with a microgram or so.

Phages f2 and Qβ have RNA as the genetic material. This is single-stranded as in all forms of life save certain plant and animal viruses. Only in viruses, some plant, some animal, some bacterial, is RNA the primary repository of genetic information rather than a temporary carrier of it. Many people have wondered whether the RNA viral genome might not serve as a retro-active messenger for the synthesis of a DNA genome to co-ordinate replication in the host cell but the evidence does not support this idea.

Although tail-less, the minute phages may have an attach-ment organ in the form of a single molecule of a different protein (the "A-protein") present in the capsid. It has been found possible to extract capsid protein, A-protein and nucleic acid from a minute phage, purify them and then remix them so that they reassemble and give rise to infectious phage again. If protein and nucleic acid will self-

Table 3.2 Properties of some prominent coliphages

Group	Phage	Nucleic acid*	Other peculiarities
Phages with contractile tails	T2	2-DNA	
	T4	2-DNA	Contain HMC
	T6	2-DNA	
Phages with non-contractile tails	T1	2-DNA	
	T3	2-DNA	
	T5	2-DNA	—— Multistage injection
	T7	2-DNA	
	Lambda	2-DNA	—— Temperate
Tail-less, 'minute' phages	ØX174	1-DNA	
	S13	1-DNA	
	f2	1-RNA	Male-specific
	Qβ	1-RNA	
Filamentous phages	f1	1-DNA	
	fd	1-DNA	

*prefix refers to strandedness

assemble *in vitro* there is no reason to suppose that any cellular conveyor belts or assembly jigs are required *in vivo*.

3.4 Filamentous phages (Figure 1.1, f1 and fd)

Another group of phages that contain single-stranded DNA are the filamentous androphages. 'Androphage' is derived from the Greek words meaning "Man-eater" and refers to the fact that they infect only male strains of bacteria. Male strains possess a long thin tube known as the "F-pilus" through which DNA is believed to be transferred to the female cell in conjugation. Filamentous androphages adsorb to the tip of this and presumably infect through it. The minute RNA phages also are male-specific but they adsorb to the sides of the F-pilus (Figure 3.8). The filamentous andro-phages are exceedingly long and slender and unique in that they do not lyse the host cell but leak out of it gradually.

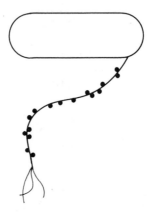

Figure 3.8 Adsorption of Androphages
A male (F$^+$) cell of *E. coli* is depicted with a single F-pilus. Three filamentous (DNA) androphages are shown adsorbed to the tip of the pilus, sixteen spherical (RNA) androphages are attached along its length. The scale is approximate.

3.5 The ultracentrifuge

Because of their small size, phages are kept in suspension by diffusion and will not sediment of their own accord no matter how long a preparation is allowed to stand. Neither can they be sedimented in ordinary bench centrifuges with top speeds in the order of 5000 r.p.m.; the centrifugal force is still not enough to overcome diffusion. Such centrifuges, however, are useful for removing bacterial cells and debris from a phage suspension. They are referred to as 'low speed centrifuges' to distinguish them from 'ultracentrifuges' in which phages can be sedimented into compact pellets using centrifugal forces of a higher *order* of magnitude.

    The centrifugal force relative to gravity (*g*) that is produced in a centrifuge is a function of its speed in revolutions per minute (n) and the radius (r), that is the distance from the centre of rotation to the point at which the force is determined. They are related by the formula

$$g = 1·12\, n^2\, r \times 10^{-5}$$

Since g increases with the square of n but only linearly with r it is clearly more efficient to construct faster rather than larger ultracentrifuges to produce higher forces. Modern ultracentrifuges have rotors with diameters in the range 17 to 27 cm and achieve forces of half-a-million g at speeds up to 75,000 r.p.m. The rotors of such centrifuges run in vacuum chambers with refrigeration to protect the specimen from the heat generated by friction.

Figure 3.9 An Ultracentrifuge
The thick steel lid has been moved aside to reveal the rotor in its armoured bowl. On the worktop stands a spare angle rotor; its cover has been removed and one tube partially withdrawn to show the construction.

The ultracentrifuge is of inestimable value in phage work, enabling phages to be washed and concentrated for analysis. Simply sedimenting phages into pellets achieves only partial purification since other matter may be co-sedimented. Better purification may be achieved by using the ultracentrifuge in one of the following more sophisticated modes, which can also yield valuable information concerning the phage's physical characteristics.

3.5.1 Rate zonal centrifugation

In this, a small volume of phage suspension is layered on top of a sucrose solution whose concentration is graded from 12.5 per cent at the top to 52.5 per cent at the bottom. During centrifugation phages travel in sharp bands at different rates according to their shape and density. Each phage can thus be separated and characterised by its sedimentation rate. Most T-phages can be banded by centrifuging for 40 minutes at 100,000 g in a 30 ml tube.

3.5.2 Isopycnic banding (equilibrium density gradient centrifugation)

If phage is suspended in caesium chloride solution of density 1·5 (0·830 g CsCl plus 1 ml of water) and centrifuged at 100,000 g for several hours the heavy caesium ions form a density gradient from about 1·4 at the top to 1·6 at the bottom and the phage forms a band corresponding to its buoyant density which is usually in the region of 1·5. The simplest method of unloading the gradient is to prick the bottom of the tube and allow the contents to drip out, collecting successive fractions. If the phage concentration is insufficient to give a visible band one may have to detect it with a U.V. monitor working at 260 nm. At even lower phage concentrations one may have to resort to infectivity titrations or the use of radio-labelled phage and scintillation counting. When the fraction containing most phage has been located its density may be determined from its refractive index using an Abbé retractometer.

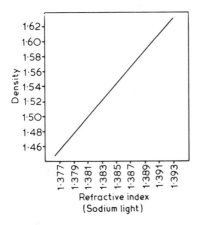

Figure 3.10 Properties of Aqueous Caesium Chloride Solutions

Grams CsCl per 100 grams of solution	Density relative to water at 20°C
32	1.3135
46	1.5158
56	1.6999

(Data from R.C. Weast's *Handbook of Physics and Chemistry*, 54th Edition published by The Chemical Rubber Co., 1973)

The decision whether to use rate zonal or equilibrium density gradient centrifugation for a particular purpose is sometimes difficult. The latter, performed as above, may require a very lengthy centrifugation for the gradient to form. This time may be reduced by using a preformed gradient. Angle rotors are intrinsically faster than their swing-out counterparts and give excellent results with caesium chloride [7]. For large scale work the high cost of caesium chloride may be prohibitive. The phage may be inactivated by either solute; T-phages survive well in both but an unknown phage needs to be tested first. A major consideration with caesium chloride is its corrosiveness. It must be rigorously excluded from any aluminium structures. In a recent paper [1] the relative merits of rate zonal centrifugation, isopycnic

banding and other procedures for the purification of phages
are critically compared. There is no generally superior
method.

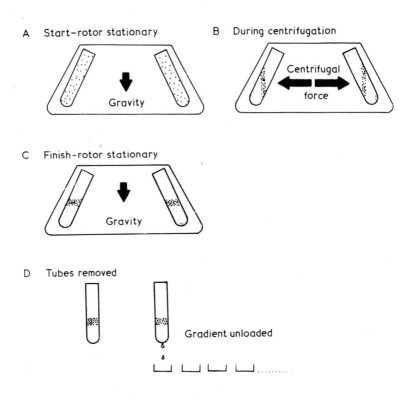

A   Start–rotor stationary   B   During centrifugation

Gravity

Centrifugal force

C   Finish–rotor stationary

Gravity

D   Tubes removed

Gradient unloaded

Figure 3.11 Isopycnic Banding of Phage in an Angle Rotor: I
A. Before centrifugation the phage is uniformly distributed in a
homogeneous solution of CsCl. During centrifugation the centrifugal force, acting in the horizontal plane, produces a lateral density
gradient of CsCl and the phage is concentrated in a band corresponding
to its own density. C. When the rotor comes to rest the gradient
re-orientates from lateral to vertical with little remixing. D. Each tube
is removed, a minute hole is made in its lower end and the gradient is
allowed to drip out slowly so that successive fractions can be collected.
Very concentrated phage is visible as a blue-brown, opalescent band.
The CsCl can be removed from the phage fraction by dialysis against
buffer.

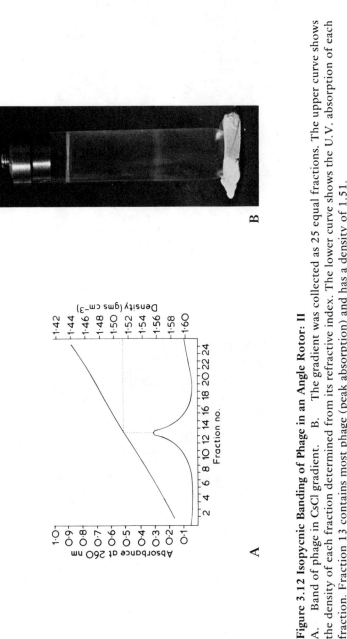

Figure 3.12 Isopycnic Banding of Phage in an Angle Rotor: II
A. Band of phage in CsCl gradient. B. The gradient was collected as 25 equal fractions. The upper curve shows the density of each fraction determined from its refractive index. The lower curve shows the U.V. absorption of each fraction. Fraction 13 contains most phage (peak absorption) and has a density of 1.51.

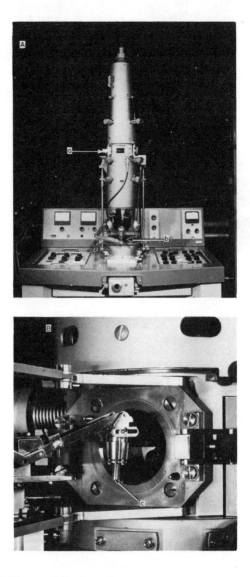

Figure 3.13 Electron Microscopes
A.   A 'conventional' electron microscope with the filament in the top
of the column, the specimen chamber behind hatch a and the viewing
screen several inches behind window b, where it can be observed
closely through the binoculars. The separate power pack and vacuum
system are not shown. B. Close-up of A's specimen chamber (opened).
A single specimen is held in the end of tube c, which is lowered
automatically into position when the hatch is closed. The vacuum
must be broken and re-established each time the specimen is changed.

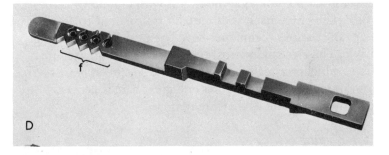

C. In this novel design the power pack and vacuum system are contained in the console; by having the column below bench level mechanical stability is improved and a very compact instrument is possible. The filament is in the lower end of the column. The specimen holder is inserted at d. The phosphor is applied directly to the back of viewing window e, giving a very bright image. It can be viewed in moderate room lighting, unlike A which must be used in almost total darkness. D. The specimen holder for C takes four grids at f. They can be viewed in any order without disturbing the vacuum. (Courtesy of A.E.I. Ltd)

3.6 The electron microscope

Phage structure can be observed only with an electron microscope. The principles of its construction closely parallel those of the light microscope but technical complexities make it a very costly instrument. Four decades of refinement have produced instruments that are not at all difficult to maintain and operate. Books on electron microscopy are numerous and manufacturers are usually willing to give technical advice so an account of the instrument itself here would hardly be germane. Where the phage-novice is more likely to need help is in the preparation of his specimen.

The electron beam has poor penetrating power, consequently the phage must be mounted on thin films of high electron-transparency, themselves supported by small copper mesh grids. The material known as 'Formvar', cast on glass, is excellent for beginners on account of its toughness. Collodion and evaporated carbon films are less grainy but more fragile. Composite films, blending the advantages of two or all of these materials can be made. The phage suspension is applied either with a spray ('atomiser') or by putting a droplet on the filmed grid and removing the excess with a point of filter paper. The phage itself is essentially transparent to an electron beam and various techniques have been used to make them visible. The electron microscope operates in high vacuum. Hence whichever technique is used the specimen must be free from water and other volatiles and great care is necessary to minimise distortion as they are removed.

3.6.1 Shadow casting

This involves the deposition, in a vacuum chamber, of a thin film of heavy atoms (gold/palladium alloy was popular) from a point source at a low angle. Only one side of the specimen will be coated for atoms *in vacuo* travel in straight lines. The accumulated metal impedes the electron beam, giving a striking 'light and shadow' effect. It was useful in early phage studies, revealing the overall shapes at magnifications of *c.* 20 000x. Since the metal actually occludes the finer details the technique is not much used nowadays except in the

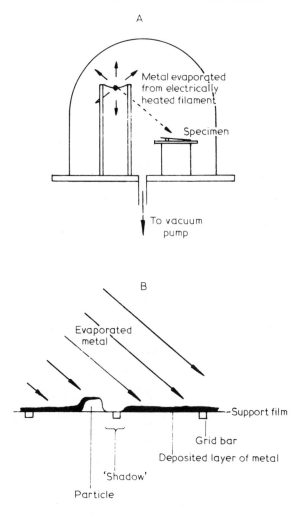

Figure 3.14 Shadow Casting
A. The specimen is tilted at an appropriate angle to the source of metal atoms. B. To show formation of the 'shadow'.

A   Phage in protein
     film on surface of
     liquid

B   Phage head
     ruptures and
     spreading protein
     extends nucleic
     acid molecule

C

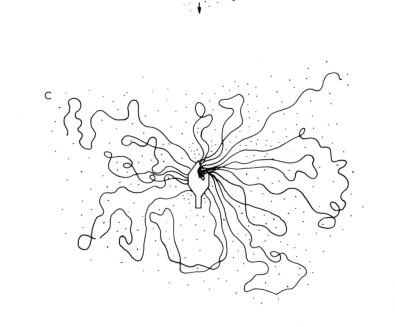

Figure 3.15 Spreading of Phage Nucleic Acid in a Protein Film
A and B show successive stages in the process leading to the fully
displayed nucleic acid molecule in C.

'freeze-fracture' process and Kleinschmidt's method for the examination of nucleic acid molecules (Figure 3.15). In the latter, phage particles are osmotically shocked in a spreading monolayer of protein (usually cytochrome C) at a water-air interface. As the protein spreads the DNA is gently dispersed over a confined region in the vicinity of the empty phage head. The spread-out molecules are picked up on a filmed grid. Platinum shadowing is applied at a very low ('grazing') angle while the specimen is rotated to achieve deposition on all sides of the molecule but not over it.

3.6.2 Negative staining

This is useful in revealing the fine details of phage structure, including the shapes of some individual monomers. The

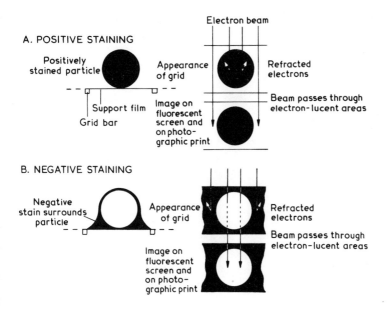

Figure 3.16 Positive and Negative Staining for Electron Microscopy
Positive staining is hardly used in bacteriophage studies except in the examination of sections of infected bacteria (*see* para. 4.7).

particles are mixed with a solution of an electron-opaque salt such as potassium phosphotungstate or uranyl acetate before application to the filmed grid. The salt does not penetrate the specimen but produces electron-dense interstices between particles and between the monomers that compose them so that a reversed image is obtained when compared to positively stained specimens.

It should be noted that shadow casting increases the apparent size of particles due to the thickness of the metal coating whereas negative staining reduces it as the stain lies above and below the real edge of the structure.

4 The latent period and the eclipse

The one-step growth experiments shown in Figure 2.4 show that phage multiplies inside the host cell and is released at the end of the latent period when it bursts, yet, if the cell is broken open, no phage can be extracted from it. What becomes of the phage during this 'eclipse phase'? The answer to this question has been obtained from three main lines of investigation. Again, the T-even phages predominate.

4.1 Electron microscopy

In ultra-thin sections of infected bacteria certain events can be discerned. Within five minutes of infection the bacterial nucleus, at first compact and discrete, breaks up into granules that migrate to the periphery of the cell and disappear. Between five and ten minutes threadlike elements appear and coalesce into spherical packets. At around the 15th minute particles resembling mature phage with head membranes and tails can be seen. If infected cells are broken open and the contents examined ten minutes or more after infection it is sometimes possible to see structures that have been called 'doughnuts' on account of their appearance and are probably head membranes containing no nucleic acid. Tail sheaths, fibres and end plates may also be seen.

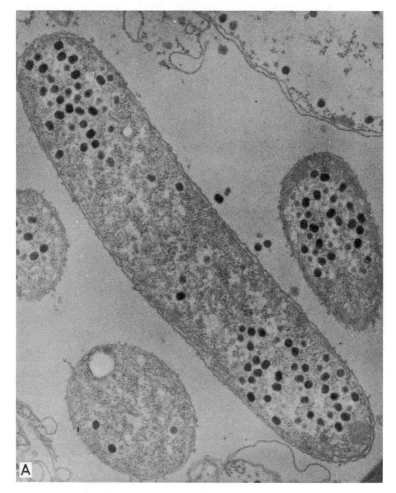

Figure 4.1 Ultrathin Sections of T4-Infected *E. Coli*
A.   Darkly staining 'condensates' can be seen. B. Many '*tau*-particles'
(arrowed) can be seen lying next to the cell membrane and a few free
in the cytoplasm. Some tubular structures (polyheads perhaps) can

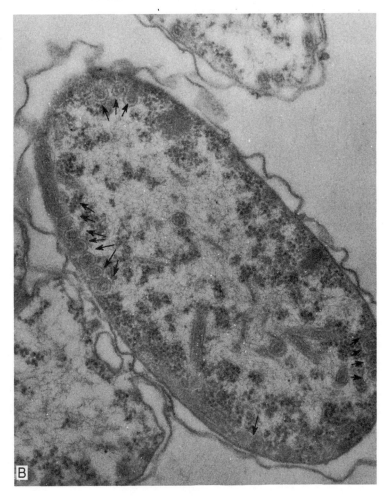

also be seen. Both the condensates and the *tau*-particles are described as 'head-related' structures but whether either or both are true precursors is not yet proved. (Courtesy of Prof. E. Kellenberger and Heidi Wunderli)

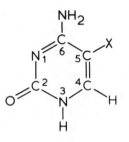

Figure 4.2 HMC
The DNA of coliphages T2, T4 and T6 contains, instead of cytosine, its analogue 5-hydroxymethyl cytosine (HMC). In cytosine X = H. In HMC X = CH_2 OH.

4.2 Biochemical analysis

Bacterial DNA contains the four nucleotide bases, adenine, thymine, guanine and cytosine found in higher organisms. Coliphage T2, however, contains hydroxymethyl cytosine (HMC) in place of cytosine and this makes its DNA readily distinguishable from that of the host cell. Thus the synthesis of the phage DNA and bacterial DNA can be studied independently in the cell. Such studies show that almost immediately on infection the synthesis of bacterial DNA ceases. The synthesis of phage DNA is detectable between the 5th and 10th minutes and continues at an almost constant rate until shortly before lysis.

4.3 Antigenic analysis

Bacteriophages are powerful antigens. The proteins that compose the various parts of the virion are all antigenically distinct and antisera may be raised against them separately if sufficient of the component can be purified. An antiserum against intact phage will contain antibodies against all the exposed protein components. The removal of any of these antibodies (but particularly those elicited by the tail proteins) will reduce its neutralising titre. Thus an antiserum against whole phage will have a certain titre but if it is allowed to react with isolated phage components that titre will be reduced. The extent of the reduction is known as the 'serum blocking power' of the preparation used. The detection of serum blocking power in a bacterial extract is therefore evidence that phage proteins are present in it. By means of this technique it was shown, for example, that head membrane antigens were present after about the tenth minute post-infection, which correlates remarkably well with the appearance of 'doughnuts' at about the same time.

4.4 The overall picture

The findings summarised in Figure 4.3 show that the phage nucleic acid injected into the cells catalyses its own reproduction and brings about the synthesis of phage proteins which are eventually assembled into mature virions. But how is this achieved? Below is an outline answer to the whole question followed by detailed consideration of two important specific aspects.

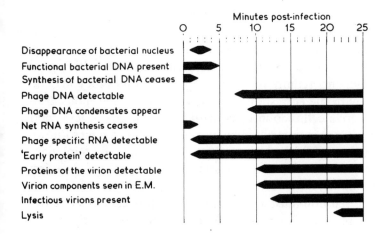

Figure 4.3 Observations on T2-Infected Cells of *E. coli* During the Latent Period
The diagram summarises the findings of several authors. Tapering of the datum bars indicates the variation in timing of each phenomenon.

The bacterial cell's own enzymes transcribe a complementary RNA copy from one strand of parts of the parental phage genome. This is called the 'early messenger'. It associates with the cell's ribosomes and directs the synthesis of the 'early proteins'. These are enzymes that (a) suppress completely, within minutes, the cell's own RNA and protein synthesis ('shut-off'), (b) degrade the bacterial DNA so that breakdown products are available for the synthesis of phage nucleic acid, (c) makes copies of the phage genome.

In the case of T-even phages, before the genome can be copied enzymes in the early protein must produce HMC, which is not present in bacterial cells. Two of the complex battery of enzymes responsible, thymidylate synthetase and deoxycytidylate hydroxymethylase, were the first enzymes to be demonstrated in the early protein and started a hunt for more that continues to the present.

Halfway through the latent period many DNA molecules, copies of the parental genome, have been formed constituting the 'replicating pool'. Units are steadily withdrawn and condensed for incorporation into virions but there is now some doubt whether the condensates seen in early electron micrographs are true precursors whereas Kellenberger's *tau* particles probably are. DNA molecules in the replicating pool are not entirely stable; breakage occurs frequently and although the fragments reunite to form complete genomes they may do so with the corresponding fragment of another molecule. Where a cell is co-infected with two or more phages genetic exchange is thus possible; indeed, specific enzymes known as 'nickase', which weakens the DNA molecule by removing a few nucleotides from one strand, and 'ligase', involved in rejoining, have been postulated. The mechanism of breakage and reunion must be a sophisticated process to ensure that the rejoined fragments make up no more or less than a complete genome.

The process of DNA replication heralds the transcription of other parts of the phage genome which specify structural proteins of the virion. These and the DNA are then assembled to form the mature, infectious virion. Finally, through the action of phage-induced lysozyme, the murein of the bacterial cell wall is digested, the weakened wall bursts and the mature particles are spilled out into the medium. Contrary to what was at one time thought, no extracellular maturation appears to be involved. Besides mature phage, lysis invariably releases phage components that, for some reason, did not become assembled ('excess antigen') and substantial amounts of free lysozyme.

The eclipse phase is of considerable taxonomic significance as it is probably the only unassailable criterion on which viruses can be defined as a group. Of all organisms, only viruses go through a stage in which they exist as nucleic acid only. However, irrefutable evidence for the existence of an eclipse phase in the life cycle of viruses other than phages is hard to obtain because of the difficulty of synchronising infection. Attempts at one-step growth experiments using plant and animal viruses have yielded results suggestive of an eclipse phase but open to dispute. With improved technique this difficulty may disappear.

4.5 Replication of the phage genome

A copy of 'replica' of the nucleic acid molecule that was introduced by the infecting phage into the host cell must be manufactured for each of the progeny. Nowadays research is concerned with the detailed biochemistry and enzymology of replication but, long before anyone knew even that the total of inheritable information ('genome') of a phage or any other organism was physically encoded in nucleic acid base sequences, S.E. Luria had analysed the overall strategy of copying. He realised that there are three fundamentally different approaches to the problem of making numerous copies of a unique original. They are shown diagrammatically in Figure 4.4. In I, where all copies are made from the original, an error would affect only that copy since no copies are made of the copy. But this method might be expected to engender severe wear and tear on the original. In II, where each successive copy is made from the preceding copy, the original is subject to minimum use but the copies will deteriorate as every blemish is cumulatively reproduced.

I   II

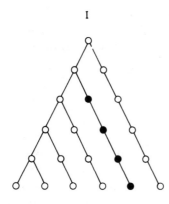 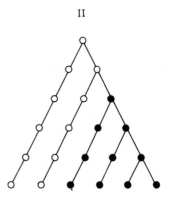

III

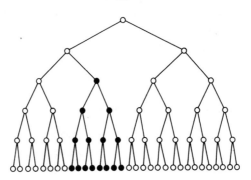

Figure 4.4 Three Strategies for Copying a Genome
Each diagram traces the lineage, through six generations, of each replica of the parental genome. At each individual replication, shown as a bifurcation, the original is placed to the *left* and the replica to the *right*. Open circles represent normal replicas, solid circles are erroneous replicas. Note the erroneous fraction:

I.   All copies are made from the original. An error in the third generation affects 1/6 individuals in the sixth generation, decreasing.

II.   Each copy is made from the preceding copy. An error in the third generation affects 2/3 individuals in the sixth generation, increasing.

III.   Both originals and copies are copied. An error in any generation affects a constant proportion of individuals in subsequent generations. Occurring in the third generation it affects 1/4.

Both I and II are inefficient processes because at any one time there is only one genome from which copies are being made. In III, where both originals and copies are copied, there is no such bottleneck. The mean quality of the copies might be expected to be worse than in I but better than in II. These ideas are not necessarily Luria's, nor does the author claim them as his own. They would probably occur to anyone faced with the task of making a large number of copies of a precious document.

But Luria took positive steps to determine which of these possible strategies represented the actual *modus operandi* of a phage. First of all he showed on paper that the distribution of mutant phages per infected cell would depend on the nature of the copying mechanism. These distributions are shown in Figure 4.5 with Roman numerals corresponding to the mechanisms in Figure 4.4. Clearly they are very different. He then set about determining the distribution experimentally. Not all errors in the phage genome copy are detectable as mutant phages but, since they occur at random, one particular mutation should reflect the behaviour of all. Luria chose to study the T2r mutant of coliphage T2, recognisable by its exceptionally large, sharp-edged plaques. He examined the bursts from a large number of individual bacteria and scored the r mutants. His findings agreed very well with the distribution predicted by mechanism III. Everything that has subsequently been learned about the T2 genome has tended to confirm that it replicates in this exponential way. The two principal papers [10], [11] in which this work is described make a fascinating study in the conjunction of theoretical and experimental analyses.

In phage ØX174, which may be typical of phages containing single-stranded DNA, the replication mechanism seems to be more like mechanism I. The original phage genome that enters the cell is single stranded and circular and is known as the 'plus' strand. Its first function is to direct the synthesis of a complementary "minus" strand and together the plus and minus strands constitute a circular, double-stranded molecule known as the 'replicative form' or 'RF'. This original RF attaches itself to a unique site on the bacterial cytoplasmic

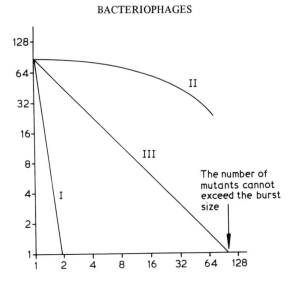

Figure 4.5 Theoretical Distribution of Mutant Phages According to the Mechanisms in Fig 4.4
Horizontal axis: number of mutant phages in the burst from a cell.
Vertical axis: number of cells yielding that number or more from their burst. The Roman numeral on each curve corresponds to the relevant mechanism in Fig 4.4.

membrane so that when it replicates the progeny RF's must remain free in the cytoplasm. There they act as master copies in that the plus strand is split away to be incorporated into a virion while simultaneously a new plus strand is formed on the minus strand. It is not yet known whether the messenger RNA that directs the synthesis of the phage proteins is transcribed from the parental or progeny RF's.

4.6 Morphogenesis of the virion

The linear order of the amino acids in a polypeptide chain, its 'primary structure', is determined by the sequence of nucleotide bases in the corresponding gene. There is a wide gap between a linear molecule and a complex three-dimensional structure like a phage but there are excellent reasons for believing that the precise size and shape of the virion, possibly

of any biological structure, are a direct consequence of the
primary structures of its protein molecules.

4.6.1 Formation of monomers

A polypeptide chain can be envisaged as a long flexible
molecule constantly changing shape as it is jostled by other
molecules in the liquid medium. If it ever does exist in that
form it must be for an infinitely short time since there are
attractive forces between different parts of the molecule
that pull it into a more compact, coil-spring shape known as
an a-helix, the 'secondary structure'. This by no means
satisfies all the residual forces on the molecule. Others, hydrogen bonding and the formation of disulphide bridges between
adjacent cysteine residues, fold the a-helix into an even more
compact 'tertiary structure' in which two or more polypeptide
chains may participate. This is the basic monomer of which
protein structures are built. Gentle dissociative chemical
treatments can be used to separate the monomers composing
some phage organelles so that their shapes can be seen in the
electron microscope. Monomers of the T-even tail sheath, for
instance, are comma-shaped. The still unsatisfied residual
forces on the surface of the monomer form a pattern which,
when aligned with a complementary pattern on another
molecule, will cause them to bind together in a highly
specific orientation.

4.6.2 Assembly

At this point the reader must permit a lapse into analogy; it
may help to clarify one of the central problems in biology.
Polypeptide chains can be regarded as different raw clays and
monomers as bricks made to a specific shape according to
the clay. Is it necessary to postulate conveyor belts to take
each brick to the appropriate place on site, bricklayers to
join them together and foremen to tell then where to start
and stop in order to produce a house of the required
dimensions? Microscopists and biochemists have rummaged
in the cell for many years to find the analogous components

but without success. Fortunately there is a very plausible explanation: these things do not exist because they are not necessary; protein monomers, unlike bricks, assemble themselves! Molecules in solution are in a state of constant rapid motion so that in an infinitesimal space of time any two molecules in a cell will come together in every possible orientation. If their surface forces are such that when they come together in one particular way they bind firmly in that position a predetermined structure will grow. The basic concept of such self-assembly has been amply confirmed by the chemical dissection and reconstitution of simple viruses including phage.

A spherical or quasispherical capsid presents no problems. Given monomers of the correct shape and bond-angle a finite regular shell will result. On the other hand a linear structure like a phage tail, growing by the addition of successive monomers to the end, is of potentially infinite length unless there is some mechanism for halting growth when the appropriate length has been reached. Figure 4.6 suggests one such mechanism suitable in principle for a structure comprising two collinear assemblies of monomers of different length, like a tail core and sheath. It works on the 'Vernier' principle. As the sheath and core grow by the addition of successive monomers, starting from a level base at the collar end, the unequal lengths ensure that the other end is never level until there are twelve sheath monomers to seven core monomers. The array of residual forces on the end is then complementary to that on an end-plate jostling nearby, which 'locks on' and thus finalises the elongation process. The degree of mismatch necessary to prevent premature adhesion need be only very slight since the residual forces on a protein molecule operate only over very short, precise distances. The 'Vernier' theory is not new and has recently been joined by others [6]. In one of these the length of the phage tail is determined by assembling it around a specific length of nucleic acid hanging out of the head. This 'tape-measure' theory has other attractions; it would explain how the nucleic acid found its way into the tail at injection. The 'cumulative strain' theory also has its adherents; in this each successive monomer must

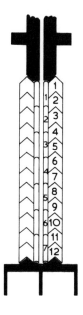

Figure 4.6 The 'Vernier' Model

be deformed a little more than its predecessor to fit the growing
tail, which terminates when the required deformation exceeds
the capacity of the monomer.

4.7 Ultramicrotomy

To observe intracellular phage development with the electron
microscope infected bacteria must be sectioned at about 100
nm thickness using an ultramicrotome. They are fixed at the
required stage in the latent period by the addition of osmic
acid (OsO_4), which also stains them, and spun down into a
pellet. The pellet is stirred with a few drops of molten 2 per

cent agar, allowed to set and cut into 1 mm cubes. A further stain, uranyl acetate, may be applied at this stage. The cubes are then dehydrated in a series of alcohol or acetone solutions of increasing concentration, ready for embedding.

The cubes of agar must be impregnated with and embedded in a synthetic resin with the appropriate cutting qualities. Methacrylate, 'Araldite' and 'Epon' have been used with success; 'Vestopal' is perhaps the most widely used for bacteriophage studies. The specimen is placed in a solution of liquid resin which diffuses into it. It is then transferred to a mixture of resin and a catalyst which when heated sets hard. This is usually done in a polythene capsule about the size and shape of a pistol bullet, with the specimen at the point. For sectioning, the point is trimmed to an approximately pyramidal shape with a flat top about one mm^2.

The knife used for cutting ultra-thin sections is normally the edge of a piece of broken glass. Diamond knives are available and are reputed to be economical in the long run if used carefully. They are not, however, an automatic choice, even for the wealthy. The coefficient of friction between the resin and the knife is important. 'Vestopal' is said to be less suitable for use with diamond than glass knives for this reason. Glass knives are made from strips of high quality glass about 25 mm wide by 6 mm thick. They are first broken across the length to make squares. The squares are then broken diagonally, thus producing four acute angles. Although the break is made by a special machine it invariably curves at the end so that only one of these angles is optimum for section cutting. A trough of liquid is fixed behind the cutting edge so that sections can float away from the edge as they are produced. They can be further stained by floating them on solutions of lead and uranium salts. Finally they are picked up on a perforated copper grid that supports them in the electron microscope.

Figure 4.7 shows the knife in its holder and the specimen in its chuck. The rest of the ultramicrotome, a miracle of ingenious design and precision engineering, is simply to ensure that they move correctly relative to each other during the cutting operation. In that, nothing short of perfection

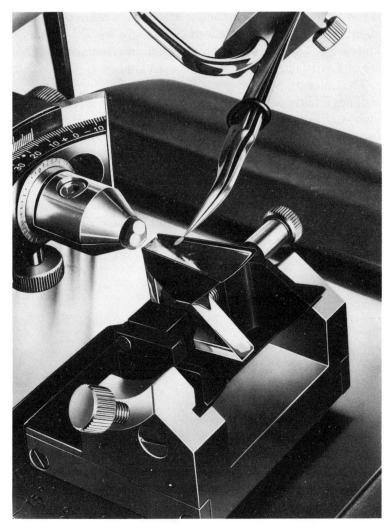

Figure 4.7 Ultramicrotomy
The supporting grid held in the forceps will be used to collect the
sections by dipping it into the liquid and raising it under them as they
float clear of the knife edge. (Courtesy of LKB Instruments Ltd.)

is good enough. With a binocular attachment magnifying about twenty times the sections can be seen coming off the knife edge. Those showing silver or gold interference colours are of a satisfactory thickness for most studies.

Ultramicrotomy is a very individual art. The foregoing outlines a fairly general technique; most workers have their own modifications in detail. Positive staining for electron microscopy is particularly idiosyncratic. No colours can be seen with an electron microscope, only light and dark according to the impedence that the specimen offers to the electron beam. Stains therefore consist of electron-dense metals like osmium, uranium and lead which, under appropriate conditions of concentration and pH, selectively react with nucleic acids or lipids and make them appear darker in the final print.

5 Lysogeny

The relationship between bacteriophages and their hosts that has been considered up to now hardly justifies the title of 'relationship' at all. It is completely one-sided in that, within minutes of their first encounter, the phage has multiplied about one hundredfold while the host is utterly destroyed. Any parasite that continues to behave in this fashion will very quickly render itself extinct, along with its host. It is not surprising, therefore, that a different sort of relationship has been found to exist, in which the destruction of the host bacterium is a rare event. It is called 'lysogeny', and the essential feature of it is that, after infection by phage, the bacterial cell continues to live and multiply normally in a manner for the most part indistinguishable from that of the uninfected cell yet harbours within it the capacity to produce a normal burst of phage under certain conditions. Such bacteria are called 'lysogenic' and the phages that produce the condition 'temperate', as opposed to the virulent phages considered hitherto.

Lysogenic cultures can be recognised by the fact that they always contain some free phage particles. The number may be very low; one per million cells or less. They can be detected by their plaque-forming ability on a different strain of bacteria. A temperate phage will not form plaques on its lysogenic host since lysogeny confers immunity to super-infection by the same phage. The sensitive strain, known as

the 'indicator' strain, may be found by trial and error, testing
a large number of strains for sensitivity to the temperate
phage. This is a tedious business. Quite possibly a high proportion of all bacterial cultures are lysogenic but the fact
has gone unrecognised for lack of suitable indicator strains.
One of the best places to look for indicator strains is in the
lysogenic culture itself. In any lysogenic culture a few
cells spontaneously 'cure' themselves of phage and thus
become sensitive to it. Ultraviolet irradiation increases the
frequency of curing.

At one time it seemed that lysogeny might be the result
of persistent re-infection by a slowly multiplying phage whose
growth rate could never catch up with that of the host so as
to infect and lyse all the cells. This situation may indeed
exist in certain 'phage-carrying' streptococci known to the
dairy industry but it was dramatically proved to be otherwise
in better known lysogenic cultures by the following experiments. Culture of lysogenic bacteria in media containing
specific antiphage serum showed conclusively that extracellular phage (which would have been inactivated by the
serum) was not essential to the maintenance of lysogeny.
Furthermore, by culturing individual cells in separate micro-
drops of medium for many generations and testing them for
lysogeny, it was shown that *all* the cells in a lysogenic culture
possessed the lysogenic property. The two experiments can
be combined, making an even more convincing demonstration.
It should have been apparent from the total immunity of the
lysogenic culture to super-infection that lysogeny affected
more than just those few cells that actually lysed. However,
it must be remembered that at the time when the concept of
lysogeny was put forward the attention of the scientific
community was focussed on the fascinating achievements of
the American workers using virulent phages. It was tempting
to dismiss lysogeny as 'experimental error' or 'misinterpreta-
tion'. or simply as a rare example of freak behaviour. It now
seems more likely that the normal phage-host relationship is
one of lysogeny and that it is the virulent phages which are
freaks.

The free phage found in a lysogenic culture arises by the

spontaneous lysis of an occasional cell, liberating a normal burst of progeny. Certain special treatments will induce all the cells to lyse, with the release of phage. The first of these to be discovered was ultraviolet irradiation. The dosage must be precisely regulated and the cells must be maintained in the appropriate physiological condition. Done correctly, irradiation results in massive, simultaneous lysis. Mutants of lysogenic cultures are known in which a small increase in incubation temperature will cause lysis to occur. Of the many chemicals found to bring about lysis of a lysogenic culture, the antibiotic mitomycin C currently enjoys the greatest popularity for experimental purposes. These treatments, physical and chemical, are known as 'inducing agents'. Their mechanisms are incompletely understood at the present time. A further form of induction, 'zygotic induction', will be described when the general properties of lysogenic cultures have been set forth in more detail. One temperate phage, coliphage P2, is said to be non-inducible; virions are released spontaneously by a small number of cells.

Every cell in the lysogenic culture can lyse and liberate phage; yet if the cells are broken open no phage can be extracted from them nor can phage be seen in sections by electron microscopy. Neither can phage proteins be detected serologically. Lysogenic cultures thus behave as if they contained a gene for phage production; a gene that is normally quiescent, functioning only occasionally as if by accident or in response to a specific stimulus. Long before the chemical nature of genes was known, the name 'prophage' was given to whatever it was that lysogenic cells contained. The prophage is transmitted to the daughter cells at division and a mathematical study of spontaneous curing suggested that prophage was present at the rate of one per cell. Phages are not very heat resistant. At 100°C their survival is measured in fractions of a second. Nevertheless the spores of a lysogenic *Bacillus* species can be boiled and still germinate to give the same lysogenic culture. Clearly, whatever mechanism protects the bacterial genes is shared by the prophage. It is now believed that the prophage does not simply *behave* like a gene; it becomes so completely integrated into the cell that it temporarily loses its separate identity and to all intents and purposes it *is* a gene. An understanding of this situation has been achieved mainly through the study of the *inheritance* of lysogeny and at this juncture readers are asked to note carefully the following caution. By far the greater part of all our knowledge of lysogeny refers to one phage, coliphage *lambda* (λ), and its relatives, the lambdoids. The published literature is vast, including A.D. Hershey's 792—page book devoted exclusively to *lambda* biology. Notwithstanding the current importance of *lambda*, one should not assume that it is typical of temperate phages in general. Too few temperate phages are known in sufficient detail for any such generalisation to be made but there certainly are some whose behaviour is very different. However, within the confines of this small volume *lambda* must provide the framework of the account, with other temperate phages being mentioned to illustrate specific points where necessary.

5.1 Inheritance of lysogeny

In bacterial matings the chromosome of the male (donor) cell passes into the female (recipient) cell via a conjugation tube (the F-pilus). Transfer proceeds linearly at a uniform rate and, by artificially interrupting the mating after a specific interval and detecting the last gene to be transferred, the linear order of genes on the chromosome can be established. Where the donor carries *lambda* the prophage is inherited like any other gene and has been assigned with precision to a position on the chromosome between the genes for galactose fermentation and tryptophan synthesis (*gal* and *trp* loci).

Where the recipient is lysogenic but the donor is not, the prophage state is maintained in the zygote. If, however, a non-lysogenic recipient receives a prophage from a lysogenic donor one of two things may happen. The zygote may simply inherit the prophage and thus become lysogenic; alternatively it may succumb to a virulent infection and lyse. This latter is the phenomenon of 'zygotic induction' and to understand it one must first appreciate how the prophage state is maintained. The prophage is the genome of the virion, comprising all those individual genes necessary for synthesis of the virion and lysis of the host cell. Why, then, do the cells not lyse? Briefly, the prophage bears also the gene for a protein called the 'repressor'. This gene is continuously active and the repressor protein associates with the prophage and prevents the other genes from being expressed. The cytoplasm of a lysogenic bacterium contains repressor which is not affected by conjugation so lysogeny is maintained in the lysogenic recipient. If however the prophage enters a cell in which there is no pre-existing repressor to bind to it, *all* its genes will be able to express themselves normally and whether lysis or lysogenization ensues depends on the relative rates of production of the repressor and the other phage proteins. The balance is delicate and the outcome depends on environmental factors.

The mechanism whereby the prophage was located on the bacterial chromosome has occupied the attention of many phage biologists. Figure 5.1 shows the principal theories that

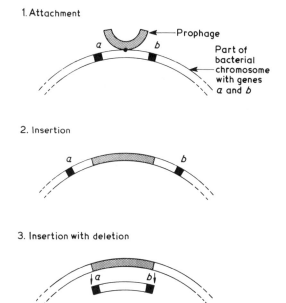

Figure 5.1 Theories of Prophage Location
For explanation see text.

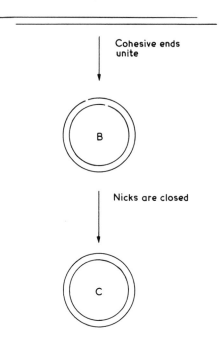

Figure 5.2 Formation of 'Hershey Circles'
DNA extracted from *lambda* virions (A) is double-stranded except at
the ends where one strand projects twelve bases beyond the other.
The two protruding sequences are on opposite strands and are complementary. If such DNA is heated to 70°C and cooled slowly the ends
join to form a helix of base pairs and the whole molecule is thus 'cir-
cularised'. At this stage (B) the ends are held by base pairing only
since no phosphodiester linkage has yet been formed to close the
'nicks' in the sugar-phosphate backbone of each strand. *In vivo* the
nicks may be closed by a specific enzyme ('ligase'), to give complete,
circular DNA (C).

have been considered. In the first (1) the prophage was
thought of as a lateral attachment to the bacterial chromosome. The idea was never popular because it was hard to
envisage how the attachment occurred and how such a
structure could replicate. The main alternatives involved the
insertion of the prophage into the bacterial chromosome,

becoming indistinguishable from it except by the nucleotide sequence. In one of these (2) the prophage was inserted at a simple break. One might think it would be easy to test such a hypothesis by seeing whether bacterial genes either side of the insertion were genetically further apart after lysogenization but it must be remembered that the prophage is only about one hundredth the size of the bacterial chromosome and genetic mapping techniques were not sensitive enough to resolve such a small separation. In the other case (3) the prophage was inserted as a replacement for a corresponding deletion of bacterial genes. If this were so, the loss of bacterial genes consequent on lysogenization should have

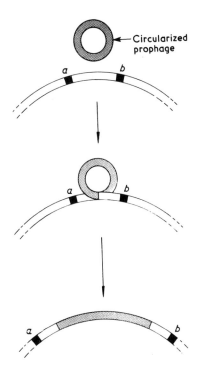

Figure 5.3 Insertion of Prophage: Campbell's Model
For explanation see text.

been detectable, but it was not. The matter was hotly debated against a growing background of biochemical and genetic evidence. It is now widely accepted that *lambda* DNA can form a circle (Figure 5.2) and be inserted into the bacterial chromosome in the manner proposed by A. Campbell [5] in 1962 (Figure 5.3), whereby there is simultaneous breakage of the prophage and bacterial chromosomes with reunion of the broken ends as shown. It is further proposed that on induction the prophage is detached by 'looping out' and 'excision' as shown in Figure 5.4. Normally the point of excision corresponds to the point of insertion but sometimes it is laterally displaced so that the phage DNA incorporates some of the bacterial genes which it can then transduce (see Chapter 7) and leaves behind some of its own so that it is defective and can no longer give rise to an infectious particle.

The phage genome can thus exist in three distinct states: a mass of compact DNA in the virion; replicating in the cytoplasm or integrated linearly into the chromosome of the host. Not all temperate phages do however; P1 prophage for instance, cannot be assigned to any particular genetic locus and is assumed to exist freely in the cytoplasm or possibly attached to the cytoplasmic membrane. It is of interest to note that no RNA phages have yet been shown to be temperate; are they intrinsically unable to lysogenize? Integration of an RNA prophage into the bacterial chromosome seems unlikely but the formation of a DNA copy of the RNA genome cannot be discounted. Host preservation by lysogenization confers a great advantage on the phage so evolutionary pressures might be expected to have selected strongly in favour of temperance in all phages. Isolation of an RNA phage and testing it for ability to lysogenize are both very laborious procedures. The real reason why no temperate RNA phages are known could be that the search has not yet been sufficiently thorough.

Reference was made earlier to the apparent normality of bacteria after lysogenization but there are exceptions. Lysogenization can convert non-toxigenic strains of *Corynebacterium diphtheriae* into toxigenic strains and alter the antigenicity of *Salmonella* strains.

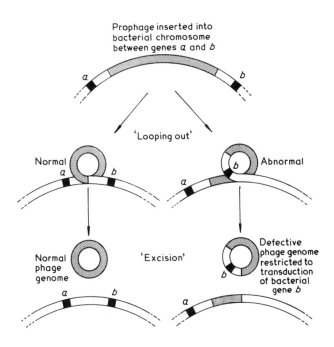

Figure 5.4 Excision of Prophage: Normal and Defective
For explanation see text.

5.2 Techniques of induction

Procedures for U.V. induction vary in detail from one worker
to the next, each finding an optimum under the particular
circumstances. The following may serve as a reference point.
To induce *lambda,* 50 ml of a dense culture is centrifuged
and the cells resuspended in 10 ml of saline. This is placed in
an uncovered glass petri dish and exposed for 20 seconds
with gentle rocking to a 15 watt U.V. lamp emitting mainly

at 254 nm from a distance of about 40 cm. The culture is
then diluted back to its original volume with broth and re-
incubated. The intensity of irradiation is affected by the age
of the lamp filament and the nature of the diluent. Small
traces of broth in the saline can absorb much of the U.V.
radiation. Lysis should occur after two to three hours. In
some phage-host systems visible lysis may not occur and to
demonstrate induction at all it may be necessary to sediment
the bacteria and plaque the supernatant on the indicator
strain.

Mitomycin C will induce many temperate phages when
added to an exponentially growing culture at a concentration
of $1\mu g \ ml^{-1}$.

5.3 Bacteriocins

A colicin is a protein produced by one strain of *E. coli* and
lethal to another strain. It may be demonstrated by the
inability of the sensitive strain to grow in broth which has
been sterilised by filtration after the colicinogenic strain has
grown in it for a short while; the inhibition is maintained even
after substantial dilution. Another way is to grow colonies of
the colicinogenic strain on agar, kill them with chloroform
vapour and overlay with agar inoculated with the sensitive
strain. On continued incubation the overlay shows growth
except over and around the colicinogenic colonies. Colicins
can be assayed by finding the highest dilution that will
produce a clear area when applied to a lawn of a sensitive
strain. This phenomenon is not confined to *E. coli*. Similar
substances are produced by other bacteria and named, likewise, according to the species. Thus *Bacillus megaterium*
produces megacins, *Pseudomonas pyocyanea* produces
pyocins and so on. Collectively these substances are known
as 'bacteriocins'.

Bacteriocins differ from ordinary antibiotics in several ways,
the most striking of which is the narrow spectrum of activity.
Whereas an antibiotic is usually active against diverse organisms
taxonomically distant from that which produces it, bacteriocins are usually active against other strains of the same species.

Electron microscopy of bacteriocins [3] reveals that there are two distinct types. There are the large bacteriocins that look like phages or parts of phages and which probably *are* phages, defective to a greater or lesser degree; then there are those which are so small that they cannot be properly resolved and whose sizes must be estimated indirectly from studies on their rates of diffusion through various media. It is to these small bacteriocins such as colicins K and V, with molecular weights in the order of 5×10^4 daltons that the rest of this account mainly relates.

One molecule of bacteriocin appears to be sufficient to bring about the death of a bacterial cell. It adsorbs to a specific receptor site on the bacterial cell wall (which may be the same as that for a phage) and remains there; whatever the influence by which it kills the cell, it is exerted from there. Treatment with trypsin can remove it and rescue the cell if applied in time. The mode of action of bacteriocins varies. Some interfere with energy-yielding metabolism, others with protein or nucleic acid synthesis; the precise mechanism is mysterious.

Whether or not a bacterium can produce bacteriocin depends on the possession of a small genetic element known as 'bacteriocinogen' ('C-factor' in the case of colicins), which is composed of DNA and can replicate autonomously. Some C-factors have been shown to pass from donor to recipient during conjugation but, since they cannot be assigned to any definite locus, they are presumed not to be integrated with the chromosome. Others actually mediate chromosomal transfer in conjugation and thus probably are integrated with it like the bacterial sex-factor.

Mere possession of a bacteriocinogen does not mean that a cell will necessarily produce bacteriocin. The medium and growth conditions have been shown to be contributory factors. The fact that ultraviolet irradiation and mitomycin C can induce colicin production and its release by the cell enhances the resemblance of bacteriocinogens to defective prophages; inevitably students ask whether that is what they are. Inducibility, the specificity of bacteriocin action, the shared receptor and the killer action all point that way but we may never know for sure. Bacteriocinogens may be defective prophages representing the degenerate remains of ancestral phages. On the other hand, if viruses are genes that acquired means of escaping from the control of the nucleus and, eventually, from the cell, bacteriocinogens may represent an early stage in their evolution.

6 Survival of phages

The phage virion is essentially a survival mechanism designed to protect the phage genome from the rigors of the environment to which it is exposed when it destroys its host and to get it into a new host when such becomes available. It is clearly efficient, otherwise phages would long ago have become extinct, yet it is common knowledge that phage preparations in the laboratory gradually lose their infectivity and eventually become useless. An understanding of the factors that affect survival of phages is thus important for anyone who intends to work with them. Furthermore, valuable knowledge of the fundamental nature of a phage may be gained by observing how it survives deliberate insults. The term 'survival' can be construed in many ways, making it important to define certain terms at the outset. Survival may imply survival as an object recognisable in the electron microscope as a phage, i.e. not utterly destroyed, but most workers take it to mean the retention of infectivity as shown by the formation of plaques or the clearing of broth cultures. A phage that has lost the ability to infect cannot be referred to as 'dead' since it is debatable whether it was ever alive. There seems to be general agreement to use the term 'inactivated' where loss of infectivity is due to a specific treatment and simply 'loss of titre' for the natural decline of infectivity during storage.

Survival of phages is usually exponential; that is to say that if the numbers of survivors are plotted on an exponential scale such as $\log_{10}$ against the duration of treatment they fall on a straight line (Figure 6.1). This is known as 'first order kinetics' and resembles the decay of a radioactive element. As with a radioactive element, survival can be expressed in terms of the half-life. For comparative purposes the inactivation rate in logarithms per minute is equally useful.

6.1 Survival in lysates

Even at room temperature many phages survive well in lysates and a working stock may often be left on the bench for a week or so with negligible loss of titre. In the refrigerator at about $+4°C$ survival is prolonged and the loss of titre may easily be less than one log per annum. Freezing and thawing a lysate usually reduces the titre to some extent but once in the frozen state, storage at $-20°C$ to $-70°C$ maintains the titre unchanged for years. Certain conditions must be observed, however, to ensure good survival. Acids produced by the host before it was lysed have a strong disinfecting action; where the host is a vigorous acid producer, a lactic streptococcus for instance, either the medium must be well buffered or the lysate neutralised for storage. Another cause of loss is the presence of unlysed bacterial debris. Phages may adsorb onto these and thus be unable to adsorb onto a new host cell. Low speed centrifugation will remove most of such debris.

6.2 Disinfection

With a few conspicuous exceptions, chemicals active against bacteria are active also against phages. Thus most general disinfectants will deal adequately with phage contamination when required. Disinfectants are very varied in their chemical make-up and modes of action. Broadly speaking, they inactivate phages by reacting with either their proteins or their nucleic acids but some react with both. Phenols and surface-active agents destroy phages by stripping off the

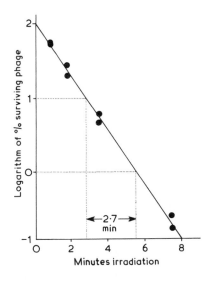

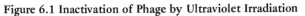

Figure 6.1 Inactivation of Phage by Ultraviolet Irradiation
A suspension of *Staphylococcus aureus* phage 42D was made in Ringer
solution and a layer 2 mm deep was irradiated in a lidless petri dish
under a U.V. source emitting 15 ergs mm^2 sec^{-1} at 254 nm. The source
was pre-run to stabilize the emission and dosage was controlled by
shielding the dish with a metal cover. Samples (0.1 ml) were taken at
intervals for plaque assay. All operations were carried out in a
darkened room to prevent photoreactivation.

The logarithm of the percentage of the original population surviving
was plotted against time and a straight line fitted by eye. This is called
the 'survival curve'. From it the time corresponding to a tenfold (one
log) decrease was found to be 2.7 minutes and hence the rate of inactivation was 1/2.7 = 0.37 logs per minute. (M. Burnett, *unpublished
data*).

protein. The various oxidising agents, hypochlorite, iodine
and hydrogen peroxide for example, denature the protein by
oxidation. Silver and mercury compounds are very active
against phages; presumably they bind to their proteins and
impair function as they do in bacteria.

Hydroxylamine and nitrous acid are chemicals that react
with nucleic acids in ways that are now well known. At low
concentrations the modification to the molecule may be so

slight that the phage is mutated but survives. At higher
concentrations the extent of the change may be incompatible
with function. Formaldehyde reacts with both nucleic acids
and proteins. Single-stranded nucleic acids are more sensitive
to it than double-stranded, a property that has been used to
distinguish them. Metabolic poisons, cyanide for instance, are
ineffective against virions as they have no metabolism to be
poisoned. So also are ether and chloroform since these substances act by disrupting lipid structures and, with the
exception of PM2, phages have none.

6.3 Experimental insults

Using dilute solutions of phenols, detergents or alkalis, the
proteins that make up the virion can be progressively dissolved
away and at any stage what remains can be examined in the
electron microscope. This 'chemical dissection' has contributed
substantially to our knowledge of bacteriophage structure.
The capsid can be ruptured osmotically to reveal the nucleic
acid; artificial stimuli such as hydrogen peroxide will make
the tail sheath contract enabling the mechanism to be studied.
These procedures inactivate the phage completely.

   Bacteriophages are about as sensitive to heat as the majority
of non-sporing bacteria. At 100°C they are inactivated almost
instantly. Between 65°C and 85°C inactivation rates can be
determined conveniently under laboratory conditions and
used to characterise phages. Below 65°C some phages are
inactivated fairly rapidly but others hardly at all. The medium
in which phages are heated has a great influence on the rate of
inactivation. Inactivation is most rapid in pure water; the
addition of salts, especially calcium or magnesium, or
proteins reduces the rate of inactivation considerably.

   The radiobiology of phages is worthy of a book to itself.
Briefly, there are two categories of radiation with which
phages may be treated experimentally: ionizing radiations
comprising principally X-rays and gamma rays, and non-ionizing
radiations comprising visible and ultraviolet light. Ionizing
radiations are highly damaging to bacteriophages, as to all
forms of life, and inactivate them by causing ionizations in

or very close to the nucleic acid molecule, resulting in strand breakage. If one of the strands of a double-stranded nucleic acid molecule is broken, the complementary strand holds the broken ends in juxtaposition, enabling the break to be repaired by specific cellular enzymes. Phages containing double-stranded nucleic acid must therefore sustain a double break to become inactivated and are thus much less sensitive to radiation than those with single-stranded nucleic acid in which every break is lethal. At one time, before the electron microscope resolved the matter, attempts were made to calculate the size of phages from their X-ray sensitivities. The principle is sound: analogously one might estimate the size of a distant bell in the dark by scattering pebbles at it and listening for the hits, but the capacity of double-stranded DNA for repair was then unknown and consequently, except in the case of phages with single-stranded nucleic acid (and that was not known at the time either) all phages appeared to be much smaller than they really are. Nor is it possible to estimate the size of the radiation target by comparing its sensitivity with that of an object of similar composition and known size such as a cell nucleus. Secondary effects whereby ionizations produce toxic substances at a distance from the sensitive site to reach it by diffusion can make the latter appear erroneously large.

The molecular mechanism whereby ionizations cause nucleic acid molecules to break is poorly understood. This complex problem has great relevance to the treatment of malignant disease with radiations and phages provide a useful experimental model.

Ultraviolet irradiation has a very subtle effect on DNA. It does not cause the strands to break. It causes covalent bonds to form, cross-linking adjacent pyrimidine bases (particularly thymine) and preventing replication. Cells contain enzymes capable of cleaving these 'dimers' and restoring function, especially if exposed to strong visible light, hence the name 'photoreactivation'. U.V. irradiated phage can also be reactivated by host cell enzymes that excise the dimer and replace it with normal bases, not always the original bases however, so that a mutation may occur. Alternatively some

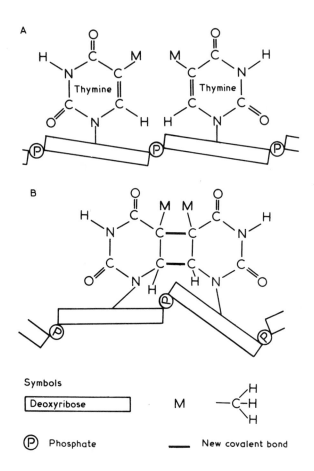

Figure 6.2 The Effect on Ultraviolet Irradiation on DNA (Dimerization)
A.   Two adjacent thymine bases on the sugar-phosphate backbone of
normal DNA.
B.   Ultraviolet irradiation has caused the formation of new covalent
bonds, creating a *cyclo*-butane ring and distorting the structure in that
area.

The exact nature of the distortion is conjectural at present. In any
case, although the diagram indicates correctly the type of bond
between atoms and groups, clear representation of accurate bond
lengths and angles in the compact double helix is not possible on a
flat page.

hosts allow the dimerized DNA to be replicated, leaving a gap in the strand corresponding to the dimer to be filled in at the next cycle of replication. This also can lead to mutation. Irradiation of phages with U.V. is a convenient method of producing mutants. The phage may be irradiated as virions in suspension or as replicating genomes inside infected bacteria. In either case a substantial proportion of those surviving a heavy dose of U.V. will be found to have mutated in some respect. If a plaque assay of survivors is performed, the curious phenomenon of 'multiplicity reactivation' may be observed. In this, the number of plaques at each successive dilution diminishes more than would be expected. The reason is that a phage may be so damaged that it cannot, even though it injects its genome into the cell, initiate a virulent infection. If the phage concentration is high, so that cells become infected with two or more damaged phages each, each phage may be able to compensate for the other's defect so that between them they cause the production of all the proteins necessary for infection and lysis.

A special form of internal irradiation known as the 'suicide technique' involves propagating the phage in a medium containing either labelled phosphorus (P^{32}) or tritiated thymidine (or other nucleoside in which H is replaced by H^3). The isotope becomes incorporated into the phage nucleic acid and, as it decays, the radiation it produces may cause breakage of the molecule. P^{32} is particularly deadly; when it is incorporated into the 'backbone' of a nucleic acid molecule each disintegration, in which phosphorus is transmuted to sulphur ensures breakage of the strand at that point. As with irradiation from external sources, phages with single-stranded nucleic acid are much more sensitive than those with double-stranded DNA.

6.4 Antisera

Phages will not infect animals but nevertheless when injected into them the proteins of the virion elicit the production of antibodies that will react with and inactivate ('neutralise') the phage. The antibodies appear in the animal's serum, which is then called an antiserum. Because of the high specificity of

the reaction of antisera with the proteins that evoke them,
antisera have played, and will continue to play, a valuable role
in bacteriophage studies. Antisera can be used to identify
phages; to inactivate a phage selectively in the presence of
bacteria or other phages; to detect the presence of free phage
proteins. It is therefore important that the budding phage
biologist should be aware at least of the way in which antisera
are made and tested; details of antibody formation and mode
of action can be found in a modern textbook of immunology
if needed.

6.4.1 Production of antisera

Rabbits are the animals of choice. Beginners may find it
easiest to inject low titre lysates subcutaneously between the
shoulder blades; 10 mls can be administered in this way to an
average-sized rabbit without distressing it. If high titre lysates
(10^{10} p.f.u. ml^{-1}) are available a faster response is obtained by
injecting a small volume (around 1 ml) into an ear vein.
Four doses, given at weekly intervals, are usually enough to
produce a powerful antiserum. About a week after the last
dose the rabbit is bled by nicking the marginal vein of the ear
and allowing about 10 ml of blood to drip into a test tube.
On standing at room temperature, the blood separates into a
clot surrounded by pale yellow serum, which can be removed
gently with a fine pipette. Anyone proposing to make antisera should first ascertain whether they are bound by any
legal prescriptions. In Britain the Cruelty to Animals Act of
1876 applies. The Home Office is usually prepared to discuss
with *bona fide* individuals the type of licence that their work
requires. Supervision by an experienced worker is essential
until skill has been attained. All materials for injection
should, of course, be free from viable bacteria.

6.4.2 Testing the potency of antisera

It is important to know the potency of any antiserum
prepared. The simple technique of demonstrating antiphage
activity shown in Figure 6.3 can be turned into a rough assay

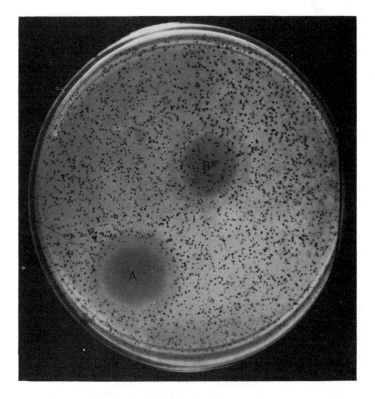

Figure 6.3 Neutralisation of Phage by Antiserum
A phage suspension was plaqued and, before incubation, a drop of
specific antiserum mixed with sloppy agar was placed on the surface
at A. A drop of control serum (from the same animal before immuni-
sation) was placed likewise at B. Under the control serum plaques have
developed normally but the antiserum has neutralised the phage and
plaque formation is completely prevented. (Courtesy of A. Qanber-
Agha)

by applying serial dilutions of the serum to the plate to
determine the 'end-point'; that which just inhibits plaque
formation. The accurate determination of the inactivation
constant (k) of an antiserum is a valuable exercise for
beginners. Dilutions of the antiserum are made in broth and
a known concentration of phage is added. The mixture is

incubated at 37°C and after a standard interval (30 minutes is often selected) the residual concentration is determined by plaque assay. The constant is then calculated thus

$$k = \frac{2 \cdot 3 \times \text{Serum Dilution Factor}}{\text{Incubation (minutes)}} \times \text{Log}_{10}\left(\frac{\text{initial p.f.u. ml}^{-1}}{\text{residual p.f.u. ml}^{-1}}\right)$$

There are other formulae for calculating k which employ *natural* logarithms. The factor of $2 \cdot 3$ is incorporated into formulae employing logarithms to base 10 (*common* logarithms) so that values of k obtained by either formula are directly comparable.

An antiserum raised against a lysate will contain, in addition to antibodies against phage proteins, antibodies against bacterial proteins as well. These latter can be removed if necessary by the addition of disintegrated bacterial cells which precipitate the unwanted antibodies as a complex that can be removed by centrifugation. The antibodies that remain will consist only of those against phage-induced proteins but not all of these, it will be remembered, are present in the virion; some are enzymic proteins involved in its production and release. Subsequent absorption of the antiserum with virions to leave only these latter is possible but technically difficult, requiring rigorous quantitative control.

Under certain conditions the neutralization of a phage by antiserum may be reversible. Papain will remove antibodies from phage and restore infectivity; so can ultrasonic treatment. Although some antigen-antibody complexes can be dissociated by dilution, the infectivity of phages can rarely be restored in this way.

6.5 Survival of phages in nature

Phages can be isolated from most natural materials in which their host bacteria are found but of the hazards they encounter there relatively little is known. The host-to-host distance is likely to be greater under natural conditions than in laboratory cultures and the demands made upon the virion by the environment correspondingly greater. A hazard likely to be encountered in Nature but rarely in the laboratory is desiccation, to which the resistance of phages is suprisingly variable, the minute RNA phages being able to survive desiccation for several days at room temperature and others being inactivated in an hour or so.

7 The importance of phage in fundamental biology

Of necessity this chapter is a long one because of the extent to which materials, techniques and concepts stemming from bacteriophage studies have permeated the science of biology in the past two decades. The chapter falls naturally into three subsections. First and foremost is the concept of the physical nature of the gene that has been provided by studies using phage. It is not true that Hershey and Chase's experiments first showed that genes were composed of DNA — Avery's work on the bacterial transforming principle *should* have convinced most people years before but it was the realisation that with phage one could introduce selected, uncluttered genes into an environment (the cell) designed for their exploitation that formed the linchpin of the genetic revolution. We must therefore take a long look at phage as a genetic system in its own right. But to what extent is genetic knowledge obtained from phage studies applicable to other organisms? In the second subsection we will consider the facile exchanges of genetic material between bacteriophages and bacterial cells that lead us to suppose that they operate on the same genetic basis. Regarding the cells of plants and animals, in which the DNA is tenaciously bound to protein and other material, we must admit that there may be profound differences yet very recently C.H. Doy and his co-

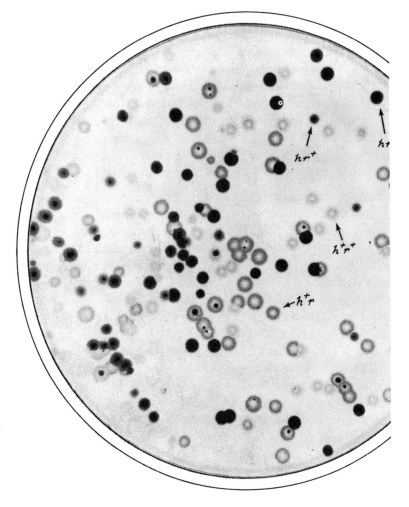

Figure 7.1 Plaque Type Mutants
Plaques formed by a mixture of T2 phages carrying mutations at the *h*
and *r* loci, plated on a mixture of *E. coli* B and *E. coli* B/2. Phages with
the h^+ allele produce turbid plaques since they lyse only *E., coli* B
while those with *h* produce clear plaques by lysing both *E. coli* strains.
Rapid lysis (*r*) mutants can be distinguished by the larger plaques they
produce. All possible combinations can be distinguished: $h^+ r^+$, h^+r,
hr^+ and *hr*. From *Molecular Biology of Bacterial Viruses* by Gunther
S. Stent: W.H. Freeman and Company. Copyright © 1963.

workers have claimed that bacterial genes conveyed to plant cells by phage would function in them and another team led by C. Merrill did likewise with animal cells. The general term 'transgenosis' has been coined to cover all such phenomena but the evidence for it is not as strong as one might wish. The third subsection comprises a selection of diverse phage-based achievements which will serve to illustrate their extent and importance.

7.1 Phage genetics and the structure of the gene

Early geneticists demonstrated the linear order of genes in higher organisms by crossing mutants and noting the degree of recombination found. At that time the gene was regarded as more or less indivisible and the establishment of the molecular basis of genetics, with the recognition that recombination can be detected *within* the gene, had to await not only the development of new techniques but also the use of a new kind of organism; one which had a minimum of genetic characters, was haploid so that no gene could escape attention through recessiveness and one in which rare events were readily detectable by the rapid screening of populations numbering billions. Bacteriophages met this requirement.

The genetics of phage, like that of all organisms, is based on the study of mutants. In outline, sensitive bacteria are simultaneously infected with two (or more) phage particles which differ in several characters and the recombinant progeny

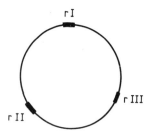

Figure 7.2 Relative Positions of the Rapid Lysis Linkage Groups on the Genetic Map of T4 Phage

counted. Since bacteriophages are haploid, lethal mutations are unsuitable for genetic studies because the phage carrying them cannot be propagated. Instead 'permissible' mutations, that is those that permit propagation, are utilised. Permissible mutations fall into two main classes: plaque-type mutations which result in the production of plaques with changed morphology and conditional lethal mutations which have a lethal effect under some circumstances ('restrictive conditions') but not under others ('permissive conditions').

7.1.1 Plaque type mutants

The most famous mutants involving plaque morphology are the rapid lysis (r) mutants of the T-even phages. Wild type (r^+) strains plated on E. coli B produce small plaques with fuzzy edges. The fuzzy edges are a consequence of 'lysis inhibition' in which reinfection with further phage of the same type delays lysis considerably so that only a fraction of the bacteria burst before the phase of active multiplication is over. If a large number of plaques is examined about one in 10 000 can be distinguished as not being restrained by lysis inhibition; these plaques, which are larger than the rest and have sharper edges, are rapid lysis mutants. There are three distinct groups of r mutation, which are designated r_I, r_{II} and r_{III} and map in different regions of the chromosome (Figure 7.2). They are distinguishable phenotypically by their

Table 7.1 Wild type (r^+) and rapid lysis (r) mutant group plaque morphology

E. coli strain used for plating	Phage group			
	r^+	r_I	r_{II}	r_{III}
B	wild	r	r	r
K	wild	r	no plaques	wild

wild = normal r^+ plaques
r = rapid lysis plaques
E. coli K is E. coli K12, lysogenic
for lambda phage

ability to plaque, and on the nature of the plaques formed, on different strains of *E. coli* (Table 7.1). Other distinct, genetically determined differences from wild type plaque morphology include those determined by turbid mutants which give plaques circumscribed by a turbid ring and 'star' mutants.

7.1.2 Conditional lethal mutants

Host range (*h*) mutant phages were used in early genetic studies. Bacteria readily acquire resistance to phage but phages often reciprocate by acquiring the ability, also through mutation, to attack these resistant bacteria. A technique for the actual isolation of host range mutants is shown in Figure 7.3. Conditionally lethal host range mutants are easily identified by plating out phage on mixed indicator strains (*E. coli* B and *E. coli* B/2 for example). Non-mutant phage lyse only their normal indicator and turbid plaques result because the other strain, resistant to the phage, remains but the *h* mutant lyses both strains of bacteria and consequently produces clear plaques (see Figure 7.1 again).

Temperature sensitive (*ts*) mutants are characterised by mutations that prevent the expression of their genes at, for example, 42°C but not 30°C. This is probably because the mutant protein is unable to retain a functional three-dimensional structure at the higher temperature. Such mutants may be isolated by plaquing at the restrictive temperature, marking the plaques that develop and re-incubating at the permissive temperature when any unmarked plaques developing should be *ts* mutants. Alternatively the cultures may be incubated at the permissive temperature first. On transfer to the restrictive temperature, wild type plaques continue to become larger while *ts* mutants remain distinctively small.

Suppressor sensitive (*SUS*) mutations of a very special kind have been described in T-even phages: these are known as 'amber' (*am*) mutations. Unlike *h* mutants, whose host range depends on whether or not they will adsorb to the bacterial strain, *am* mutants will adsorb to both the permissive

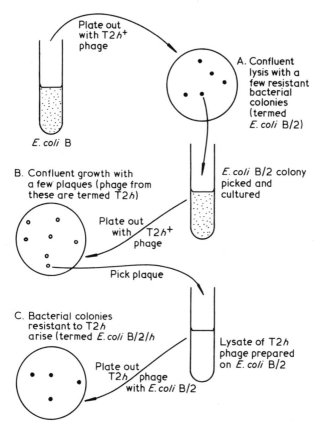

Figure 7.3 Scheme for the Isolation of Phage Resistant Bacteria and of Host Range Phage Mutants

Confluent lysis results when *E. coli* B is plated (after allowing time for adsorption) with a gross excess of wild-type (h^+) T2 phage (A). A few resistant bacterial colonies will be present, these are termed B/2 bacteria, that is resistant to T2h^+ phage. If an excess of T2h^+ phage is plated with B/2 bacteria no confluent lysis is found but a few plaques arise from infection by mutant phage particles designated T2h (B). Then if *E. coli* B/2 and phage T2h are plated together, B/2/h bacterial mutants can be obtained resistant to both h^+ and h phage types (C). and so on.

and restrictive host. They are characterised by a mutation generating the 'amber' nonsense codon (UAG) in a gene with a vital function so that they can multiply only in bacterial strains carrying a suppressor of this codon. It is said that 'amber' is the English equivalent of 'Bernstein', the name of someone associated with the discovery of these mutants, and has no more significance than that. The mutants themselves are of the greatest significance because they are readily isolated and have helped fill in gaps all round the genetic map where no other mutants were available.

7.1.3 The technique of genetic analysis by crossing phage mutants

Once a series of mutants has been isolated one can begin to construct a genetic map for the region to which the mutants pertain. In principle, bacteria are mixedly infected with two mutant phage types and the percentage of wild type progeny that result, if any, is recorded. For such recombination experiments at least three particles of each mutant phage are added per bacterial cell so that it stands a good chance of being infected by both. Most of the progeny of such a 'cross' will resemble one or other of the parents but in a few the genetic information may have been recombined to give a genome without either mutation, *i.e.* the wild type. The further apart the mutational sites lie, it is assumed, the greater is the chance of recombination occurring between them so, by converse reasoning, the less frequent the wild type recombinant the closer the mutational sites. Where *no* wild type recombinants occur it is taken that the mutational sites are identical, at least within the limits of resolution of the technique.

It must be remembered that with phage one is dealing with a dynamic system in which recombination occurs randomly within a multiplying population of phage chromosomes, each member of which can partake in a succession of recombinational events. Under such circumstances one cannot hope to discover the outcome of any single mating event and the only reasonable approach is to regard phage recombination as a problem in population genetics which requires mathematical treatment for its solution. Besides *recombination,* mixed infection may also result in *complementation,* where both mutants reproduce in a cell where neither could on its own, and *phenotypic mixing,* where progeny virions consist of the DNA of one mutant and the proteins of another. Only recombination, however, produces genetically stable wild type progeny.

7.1.4 Fine structure analysis of the r_{II} region

The first system of genetic fine structure analysis was evolved by S. Benzer in the mid 1950's. The development of this system was a direct consequence of his observation that, in general, r_{II} mutants of T4 phage do not plaque on *E. coli* K, although they infect and kill this strain. To remind you, r_{II} mutants are easily recognised since they produce large, distinctive plaques on *E. coli* B but do not normally plaque on *E. coli* K. However, an r_{II} mutant can grow on *E. coli* K if the cell is co-infected with a wild type phage — it is as if the wild type makes good some deficiency of the mutant. Benzer's success in exploiting the r_{II} system of phage T4 depended to a very large extent on the development of several distinct types of mapping procedure which reduced the work load enormously. One was the use of deletion mutants to order point mutations; another was the complementation or 'cis-trans' test which led to the recognition of units of genetic function.

7.1.5 The use of deletion mutants to order point mutations

Deletion mutants show no tendency to revert spontaneously to the wild type and include those mutants in which large deletions or alterations of the chromosome have occurred. Point mutations, on the other hand, revert spontaneously at measurable rates and behave as if their alterations were localised at single points and are correctable. Conceivably they arise when the DNA molecule of the phage undergoes alteration of a single base pair; often this substitution can be reversed with relative ease. Point mutations, not deletion

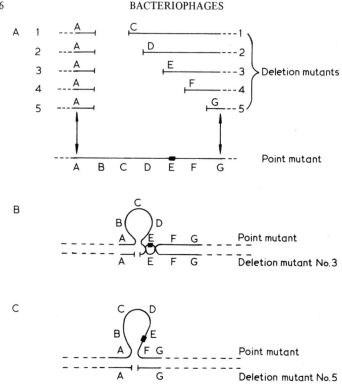

Figure 7.4 The use of Deletion Mutants in Genetic Mapping
A.   A series of deletion mutants with the extent of their deletions
indicated, and the relative position of a mutational site in a point
mutant. B.   Pairing of chromosomes carrying point mutation with
deletion mutant No. 3. Wild type recombinants can result. C. Pairing of
chromosomes carrying point mutation with deletion mutant No. 5. No
wild type recombinants can result since if recombination does occur
with the deletion mutation the point mutation will not be involved.
Hence wild type recombinants may be formed when the point mutant
is crossed with deletion mutants 1, 2 and 3 but not with 4 and 5.

mutations, are the ones that must be mapped if the fine
details of genetic structure are to be probed. However, to test
thousands of point mutations against one another for
recombination in all possible pairs would require millions of
crosses and is obviously impracticable. Benzer's very neat

solution to this problem was to use deletion mutants to assign the point mutations to certain regions of the chromosome and thus considerably reduce the requirement for carrying out pairwise crosses with point mutations. In essence, recombination tests are made between two mutants; one with a point mutation, the other with a deletion. A negative result, that is no wild type recombinant progeny, indicates that the deletion overlaps the point mutation; a positive result indicates that it does *not* overlap (Figure 7.4).

For his analysis Benzer isolated large numbers of point and deletion mutants in the r_{II} region and allocated each point mutation to one of seven segments of the region by means of a set of deletion mutants as shown. Then he further localised the point mutations into one or other of 47 segments using another series of deletion mutants which intruded into the deletion segments of those previously used. Mutants falling into each of these small segments were sometimes further localised by more deletion mutants and were then mapped by pairwise crossing. Those point mutants that showed recombination were concluded to be at different sites; the order of sites within a segment can be established by making quantitative measurements of the recombination frequencies with respect to one another and ordering the point mutations accordingly at appropriate distances apart on the chromosome.

7.1.6 The limits of genetic fine structure analysis

Down to what level can the gene be divided? Can it, for example, be mapped at the level of individual nucleotides or is the region over which recombination takes place much larger than that?

When Benzer crossed members of his r_{II} point mutation population and selectively scored for the frequency of wild type (r^+) recombinants he was able to detect very rare recombinants between adjacent sites; theoretically those occurring at frequencies of 0.0001 per cent. He found, however, that the lowest frequency at which r^+ recombinants were detected was 0.01 per cent even though the method was a hundred times more sensitive than this. Hence Benzer seemed to have

discovered the minimum distance between two genetic sites
still separable by recombination. His figures indicated a
physical distance corresponding to about three nucleotides
(one codon) as the minimum recombinational unit but recent
work has shown that this is not quite true; recombination
within a codon can occur but is relatively difficult to detect.

7.1.7 Autonomous functional units defined by the complementation or 'cis-trans' test (Figure 7.5)

The hereditary structure needed by the phage in order to
multiply consists of many parts, distinguishable by mutation
and recombination. But, one may ask, is the r_{II} region, because
it controls one characteristic, namely whether or not the phage
can plaque on E. coli K, to be thought of as one or hundreds
of genes? Although one function is involved it could be argued
that growth in E. coli K requires a series of biochemical
reactions, each controlled by a different portion (gene?) of
the r_{II} region. The absence of any one step would block the
final result, the ability to plaque.

Mixed infection of E. coli K by wild type phage and a
phage carrying two r_{II} mutations still results in the formation
of plaques containing both wild type and doubly mutant
phages. The wild type phage is able to promote the growth
of the r_{II} mutant by making good both functional defects. On
the other hand, when E. coli K is mixedly infected with various
combinations of r_{II} mutants in what is called a 'trans' arrangement: (r_{II} $a^- b^+$) or (r_{II} $a^+ b^-$), neither of which yield plaques
on this host when plated alone, either no plaques or normal,
wild type plaques develop. The plaques that do arise are a
result of complementation, not recombination, since the
phage particles in them usually consist of the two mutant
types, not the wild type.

Using this knowledge, Benzer devised what is known as the
'complementation test' which measures the ability of mutants
to complement each other, that is each make good the other's
defects. With this test he classified r mutants into two clear-
cut functional groups, A and B, each of which can function

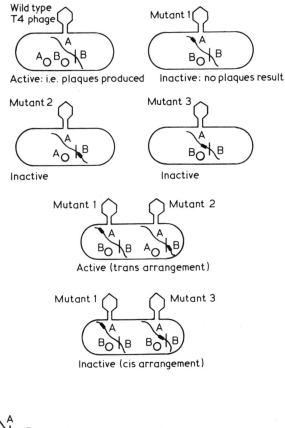

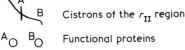

Cistrons of the r_{II} region

A_O B_O Functional proteins

Figure 7.5 The Complementation or Cis-Trans Test
Wild type cistrons produce wild type functional proteins while certain
mutant cistrons cannot. Therefore T4 is active in *E. coli* K only if both
A and B cistrons are functional in an infected cell. In this Figure *E.
coli* K is infected by T4 phages; either the wild type or mutants carrying gene mutations in the r_{II} region which result in the production of
non-functional protein in either the A or the B cistron. A and B
proteins result only when both cistrons, either on the same or separate
phage genomes, are present in a functional state in any one cell. (only
the r_{II} region of the phage genome is indicated)

independently of the other. An r mutant in the A region can complement an r_{II} mutant in the B region but two in the A or two in the B region ('cis' arrangements) cannot do so. Hence both A and B represent a single functional unit. These units were termed 'cistrons'; it can be said that the r_{II} region is composed of two cistrons, the A cistron and the B cistron. The approximate genetic lengths of these cistrons can be inferred from the recombination frequencies of mutant loci identified at their extreme ends, the A cistron being about 6 map units long and the B cistron about 4 map units long. A map unit is the genetic distance between two loci that recombine with a frequency of 1 per cent. This work gave a new perspective to genetics and it seemed that 'cistron' might replace the old term 'gene'. Habits die hard, however, and geneticists still refer to the unit of function as a 'gene' although they may think of it as a cistron.

7.1.8 Complete mapping

The r_{II} region, now mapped in considerable detail, is but a small part of the T4 chromosome. A genetic map exists for the whole of the T4 chromosome, albeit in somewhat less detail. It was not all arrived at in the same manner as the r_{II} region map.

The bulk of the mapping, in fact, was accomplished using *ts* and *am* mutants. R.S. Edgar and R.H. Epstein used Benzer's cis-trans test to assign more than 1000 of these mutants to different genes and some 60 genes had been identified and mapped by the mid 1960's. Edgar, working with *ts* mutants, would mixedly infect *E. coli* with two of them and maintain it at a restrictive temperature. If the two mutants complemented each other to produce lysis they were assigned to different genes; if they did not so complement each other they were assigned to the same gene. Epstein similarly tested his *am* mutants, infecting restrictive *E. coli* with them in pairs and noting whether lysis occurred. Finally the outcome of mixed infection of non-permissive *E. coli* with a *ts* and an *am* mutant allowed determination of their mutual complementation capacity and hence the relative

Figure 7.6 The Genetic Map of Bacteriophage T4
Each small division of the inner circle represents 10^3 nucleotide pairs
— the total length of the T4 DNA molecule determined by electron
microscopy is about 1.66×10^5 nucleotide pairs so the agreement is
close. Gene positions are indicated by lines or stippled bars and
functions of essential genes can be inferred from the consequences of
the corresponding gene defects, shown in the boxes. Functions or
designations of non-essential genes (all others) are defined on the outside of the circle for example, ac: acriflavine resistance; imm: superinfection immunity. (Reproduced by kind permission of W.B. Wood
from *Handbook of Genetics,* Vol. 1, edited by R.C. King, Plenum,
New York, 1974)

gene assignment of the two mutant types.

This mapping demonstrated that the entire chromosome of T4 is a single unit and is genetically circular. By examining the contents of restrictedly infected cells biochemically or in the electron microscope it was possible to determine the nature of some of the restrictive lesions. These are shown in the boxes on the map (Figure 7.6); for a full explanation of all the symbols the reader must consult Professor Wood's article. Genes relating to the same or associated functions are seen to be clustered but, and the significance of this is considered more fully in the section on 'Control Mechanisms', clusters relating to similar functions may be found on widely separated segments of the map.

Another phage well characterised genetically is *lambda;* special attention has been centered on the nature of the link between the prophage and the bacterial chromosome. Early genetic studies were carried out in the 1950's by F. Jacob and E.L. Wollman using characters such as host range, plaque size and plaque morphology. Other mutant types studied more recently include the *c* mutants which are fairly readily isolated and recognised. They interfere with the ability of the phage to lysogenise so that clear, instead of turbid plaques result on indicator strains (*not* mixed). These mutants fall into three functional groups, c_1, c_2 and c_3, each of which is now known to be a cistron. More recently A. Campbell has isolated and mapped a series of amber mutants of *lambda* which have defined the genetic map (and the behaviour of prophage) more fully. A partial genetic and functional map is given in Figure 7.7

7.2 Transduction

The importance of bacteriophages in genetics is twofold: besides providing a model genetic system for study in their own right, they can mediate the transfer of *bacterial* genes from one cell to another by the process of 'transduction', which literally means 'leading across'. Transduction was first noted in 1951 by Lederberg, Lederberg, Zinder and Lively who were trying to repeat with *Salmonella typhimurium* experiments that had previously been used to demon-

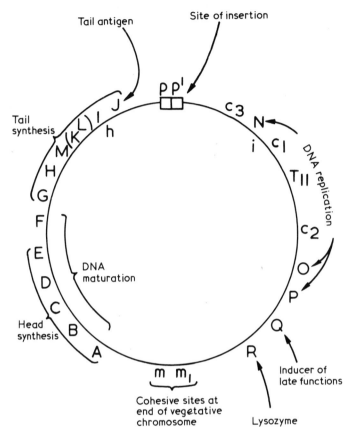

Figure 7.7 A Partial Genetic and Functional Map of *Lambda* Phage
Genes c_1, c_2 and c_3 are concerned with the establishment of prophage
and c_1 is also concerned with the maintenance of lysogeny; h and i
determine host range and immunity respectively; the capital letters
represent sites of genes based on the analysis of amber mutants by
Campbell.

strate conjugation in *E. coli* K12. When they mixed two
auxotrophic strains of *S. typhimurium,* unable individually
to grow on minimal medium, prototrophic recombinants
able to grow on minimal medium arose. However recombinants resulted not only when cultures of the two *Salmonella*
strains were mixed but also when a culture of one strain was
treated with a cell-free filtrate of the other. This was in

distinct contrast to conjugation where cell-to-cell contact is
necessary. Zinder and J. Lederberg then turned their
attention to the nature of the agent in the filtrate and soon
recognised it as a phage, named in this instance 'P22'. Transduction has since been reported in genera as diverse as
*Bacillus, Escherichia, Proteus, Pseudomonas, Shigella and
Staphylococcus.*

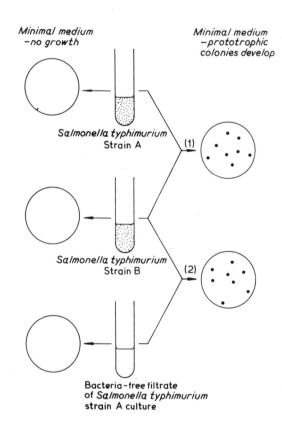

**Figure 7.8 Procedure Adopted by Lederberg, Lederberg, Zinder and
Lively to Demonstrate the Process of Transduction, as Distinct from
Conjugation, in** *Salmonella typhimurium*
Cultures are plated either singly or in various combinations on minimal
medium. The prototrophic colonies arising from combination (1) *could*
be due to a process similar to conjugation; combination (2) shows that
this is not so.

7.2.1 The mechanism of transduction

When a phage lysate is prepared from a donor bacterium only a small proportion of the progeny particles (10^{-5} to 10^{-7}) can transmit genetic characteristics to a subsequent host. These transducing phages carry fragments of donor chromosome in place of their own viral genes. Hence these phage particles are defective; they cannot reproduce themselves.

Often transduction can occur for *any* genetic markers of the donor bacterium. However, since the size of the DNA fragment that can fit inside a phage head is limited, only closely linked markers are co-transduced by the same particle. The amount of DNA in the viral genome is known to be about one hundredth of that in the *E. coli* or *Salmonella* chromosome and consequently it seems reasonable to suppose that the phage cannot transduce donor fragments appreciably larger than this. Studies on the *thr-leu-azi* region of the bacterial genome of *E. coli* K12 have revealed that the *thr-leu* portion represents the extreme length of donor fragment which can be transduced by P1 phage; this portion constitutes about one fiftieth of the *E. coli* genome.

7.2.2 Types of transducing phage

There are two types of transduction; generalised and restricted. In generalized transduction any small region of the donor chromosome can be transduced as in the P22/ *S. typhimurium* system outlined earlier. Restricted transduction on the other hand is limited to phage that can lysogenise bacteria; transfer of chromosomal genes is restricted to a cluster of genes adjacent to the prophage location on the bacterial chromosome. Restricted transduction, therefore, is possible only with lysates produced by induction of a prophage and not (in contrast to generalised transduction) with lysates produced by lytic infection. A good example of a restricted transducing phage is *lambda*. On induction *lambda* prophage will occasionally undergo exchange with the nearby *gal* segment of the bacterial DNA and particles defective as phage but able to transduce the *gal* genes will result. These particles are called λ*dg* (for *lambda,* defective, *gal*) and constitute only a minute

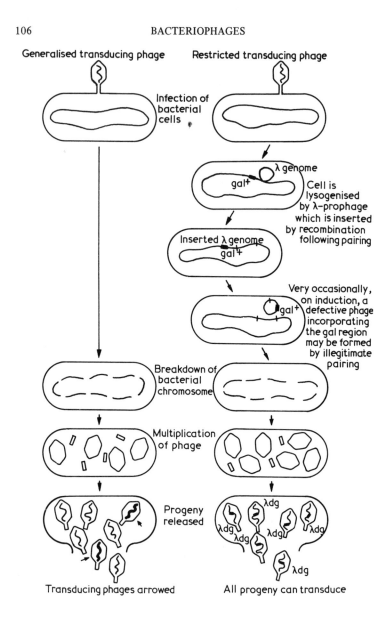

Figure 7.9 Production of Transducing Phage
In the production of generalised transducing phage a proportion of the
phage produced by a cell may contain a fragment of the bacterial
genome and hence be able to transduce that character. Restricted
transducing phage results when the cell is lysogenised by λ prophage

inserted into the chromosome; on induction an occasional prophage exercises abnormally, incorporating the bacterial gal^+ gene. It replicates, and all the progeny in that cell can transduce gal but, since such abnormal excision is rare in a population, an LFT lysate results.

fraction of particles in the lysate, which is called an LFT (low frequency transduction) lysate

7.2.3 Partial diploids

If a culture of a gal^- mutant of *E. coli* K12 (one which cannot utilise galactose), which may be either sensitive to or lysogenic for *lambda*, is infected at high multiplicity with an LFT preparation of λ*dg* phage obtained by induction of a lysogenic gal^+ strain, clones of gal$^+$ bacteria can be recognised after growth on agar containing galactose together with suitable indicator dyes. The majority of these gal^+ transductants differ from the original gal^+ donor bacteria in two respects. Firstly they are unstable for the gal character and gal^- progeny are segregated with a probability of 10^{-3}/cell/generation. Transductants must therefore be diploid for the gal region since the gal^- gene must have persisted in order for it to segregate later. Hence the transduced marker does not replace its recipient allele but is added to the recipient genome. Partial diploids of this sort are called 'heterogenotes'; they are usually doubly lysogenic, containing both λ*dg* and regular *lambda*, and are unstable for the λ*dg* region. Secondly, when such cells are induced by ultraviolet light, they yield a high frequency transduction (HFT) lysate, in which about half the particles are λ*dg* and half are normal phage. Regular *lambda* has to be present in these cells to act as a helper, providing the functions of genes missing in λ*dg*.

7.2.4 Complete and abortive transduction (Figures 7.10 and 7.11)

Transduction is said to be complete when the fragment of donor bacterial chromosome (called an 'exogenote') introduced by the phage is inserted, by recombination, into the chromosome of the recipient cell. The inserted markers are then

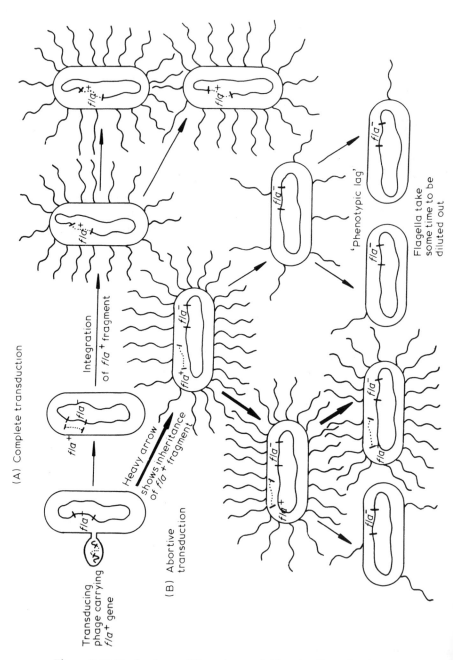

Figure 7.10 Mechanisms of Complete and Abortive Transduction
In complete transduction (A) the *fla*⁺ fragment is incorporated and

replicates so that all progeny are equally motile. In abortive transduction (B) the fragment is inherited unilinearly; in those progeny
that receive it flagella are fully developed, in those that do not the
character is gradually diluted out.

transmitted regularly to all progeny. Sometimes however, the
exogenote does not become inserted into the bacterial
chromosome, does not replicate and is transmitted unilinearly
so that at any one time only a single cell of the clone of cells
produced from the transduced cell possesses the fragment.
This phenomenon is known as 'abortive' transduction. The
exogenote still expresses the function of the gene or genes it
possesses and it follows that the gene products (enzymes or
enzyme-forming systems) are synthesised in every cell through
which it passes. These products may be transferred cytoplas-
mically; derivatives of the cell may go on dividing for a
limited number of times until the gene products have been
diluted out, exhibiting what is known as 'phenotypic lag'.

Abortive transductants for a nutritional marker form
minute colonies on minimal medium; these contrast with the
larger colonies of complete transductants. On the other hand,
when genes concerned with motility are transduced complete
transductants yield swarms of growth in soft gelatin agar
while abortive transductants give trails of colonies leading out
from the point of inoculation.

Transduction has proved an extremely valuable tool in
mapping the bacterial chromosome: the degree of linkage of
genetic markers can be inferred from the relative frequency
with which they are co-transduced; the greater the frequency
of co-transduction the closer the linkage. This enables the
bacterial chromosome to be mapped in finer detail than is
possible with conjugation experiments. In the latter, the
position of any particular gene can be determined with an
accuracy of about 1 per cent of the total length of the
bacterial chromosome. Transduction allows the position of
a gene to be determined within a 1 per cent fragment so that
the two methods are complementary. The loci determining
tryptophan synthesis in *Salmonella* were mapped using the
transducing phage P22 as genetic vector, while the pathway

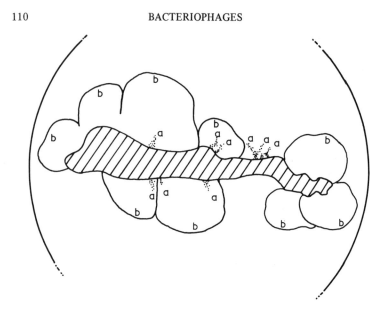

Figure 7.11 Transduction in *Salmonella typhimurium*
A non-motile strain (*fla⁻*) of *S. typhimurium* was treated with phage
lysate from *S. typhimurium* LT2, a flagellated (*fla⁺*) strain, and
inoculated in a broad streak (hatched) on a plate of semi-solid gelatin
medium. After 24 hours incubation at 37°C trails of small colonies of
abortive transductants (a) could be seen emanating from the inoculum
together with *fla⁺* swarms (b) each derived from one completely
transduced cell. (Drawn from a photograph by Dr. S.E. Smith)

of L-arabinose utilisation in *E. coli* was shown, using mutant
E. coli strains and a transducing phage, to be mediated by
three genes, each specifying one enzyme, and a regulator gene
whose function was thereby defined.

   Whether or not transduction occurs to a significant extent
in Nature is a matter for conjecture. It is widely argued that
bacteria tolerate phages because transduction confers evolutionary flexibility. Restricted transduction would seem to
give the bacterium a small return for harbouring the seeds of
its own destruction. The fact that phages and their hosts, both
lysogenic and sensitive, can be isolated from the same environ-
ments suggests that some delicate equilibrium exists in which
each is necessary for the other's survival.

7.3 The wider aspects: A miscellany

7.3.1 Control mechanisms

In the 1930's and 40's Energy and Phosphorylation were the
centre of biological attention. Through the 50's and 60's,
DNA and the Genetic Code reigned supreme. Now emphasis
is shifting to Control, with the ultimate objective of understanding Cell Differentiation and Morphogenesis. Somewhere
over the horizon, the other side of the millenium probably,
looms comprehension of the Brain and Consciousness — too
far away to concern us here. The basic problem of control is
as follows. At any particular time an organism may require
only part of its genome to function and different parts may
be required to function in a programmed sequence as, for
example, when a cell goes through the complexities of
division. If the organism is multicellular, different parts of
the genome may be required to function depending on the
differentiation of the cells, yet all the cells contain the same
genome, derived by successive replication from that of their
original unicell (spore, zygote, egg. . .). So how are the required
parts of it promoted and the remainder suppressed in the
appropriate areas of space and time? The control systems that
undoubtedly exist in higher organisms resist experimental
analysis because of their size and complexity. Even bacteria
have proved to be less amenable than was originally thought.
What more natural then, than to seek control systems in those
smallest and simplest of self-replicating entities, the phages,
and try to understand these. When something has been
accomplished in this direction, pro- and eu-karyotic cells
might be tackled with more élan.

The very speed of phage multiplication and the fact that
virion proteins are produced in approximately the proportions
required are advanced as indications that control systems
operate generally. The classification of virus-induced proteins
into 'early' and 'late' supports this. The case of lysozyme,
however, is ambiguous. Clearly it must *function* last of all,
otherwise unassembled phage components would be spilled
out prematurely into the medium but that does not mean that

it is necessarily synthesised last. A cell may contain a high concentration of lysozyme yet fail to lyse. A very plausible explanation (the 'equilibrium hypothesis') is that lysozyme is synthesised well before the end of the developmental process and it begins to dissolve away the cell wall at once but that repair is continuously effected until the cell's capacity for re-synthesis is reduced to a critical level as a result of the infection and lysis then ensues.

Some putative control mechanisms are analysed in Table 7.2. Positive control implies that a gene cannot be expressed *until* the product of another gene is present; negative control

Table 7.2
Control mechanisms in phage development: some presumptive examples

Level of control	Type of control	Example
Injection	Multi-stage	T5
Replication of the genome	Negative	P2
Transcription	Positive	Gene 1 of phage T7
	Negative	*Lambda* repressor
Translation	Negative	RNA-phage replicase
Gene-product function	Dynamic equilibrium	Lysozyme

means that one gene is not expressed *because* the product of another gene is present. Gene c_1 of phage *lambda* produces a protein (the 'repressor') that binds specifically at two points on the prophage and prevents transcription of all the other genes. This is perhaps the best known example of negative control. In RNA phages the accumulation of coat protein inhibits the translation of the replicase gene. Gene 1 of phage T7 produces a protein that is thought to bind to RNA polymerase and alter its specificity so that it no longer transcribes early genes but transcribes late genes instead. This can be regarded as an example of positive control. Given that positive and negative control are both possible, one can envisage a general "cascade" mechanism of regulation in which a gene product specifically shuts off genes no longer required and promotes the next in sequence when its concentration reaches a certain level. Evidence for other positive and negative controls is gradually accumulating but R. Calendar's recent review of progress in this area [4] reveals that no comprehensive control scheme is yet in sight for any phage.

At one time it seemed possible that the order of gene expression might be a reflection of their linear order in the nucleic acid molecule, which passed through some cellular organelle that read it as a computer reads a tape, with appropriate pauses. This attractive hypothesis is seemingly contradicted by the genetic map of phage T4 (Figure 7.6) on which genes for related functions are often very far apart and which bears no relation to the developmental sequence in any circular permutation. Thus the cascade mechanism depending not on linear order but on specific product binding to other genes reached by diffusion, is now a more appealing model. There is some evidence, however, that linear order may be important in an entirely different mechanism in phage T5. This phage injects its genome, not as one piece but as two (or more?) pieces separated by a delay. This could be a form of control which limits injection to those genes required at the time, leaving those not yet needed in the phage head where they cannot function.

In at least one case, replication of the genome seems to be under some sort of negative control. Superinfection of a lysogenic bacterium which carries P2 prophage with a virulent P2 mutant yields only the virulent phage. This is not a consequence of prophage attachment because if the cell is superinfected simultaneously with the virulent mutant and another phage of the prophage type, again only the virulent mutant is produced, i.e. the superinfecting temperate phage is blocked as completely as its prophage. Hence all the positive elements must be present, otherwise the virulent mutant would not grow, and control of the temperate phage must be negative.

7.3.2 The Qβ equation

Conventional molecular biology assumed that all genetic information was primarily stored as DNA, transcribed into RNA and thence translated into protein so the existence of viruses with an RNA genome was rather embarrassing at first but has since turned out to be a blessing in disguise. A phage genome consisting of a single RNA molecule can often be extracted in one piece, and used to prime the synthesis of all the proteins that it is known to specify in cell-free extracts of *E. coli*. It thus earns the title of 'polycistronic messenger' and, more than any other natural messenger, has been used to study details of translation such as polarity (from which end does translation commence?) and termination (how is the complete specification cut into one-protein pieces?). These matters have been extensively reviewed [9].

Phage Qβ is a favourite experimental material on account of the stability of its replicase. There appear to be only three Qβ proteins: the coat protein and the A-protein in the virion and the replicase in the infected cell. The RNA molecule in the virion is known to contain about 3300 nucleotides, sufficient to code for 1100 amino acids. The coat protein monomer consists of 129 amino acids and the A-protein accounts for 320 or so more, leaving 651 for the replicase. Pure (as opposed to 'purified') replicase is not yet available but first estimates suggest that it has at least twice that number. There *may* be other Qβ proteins as yet undiscovered and equally there *may* be RNA sequences that serve some purpose other than to be translated into protein but the over-all impression gained from these figures is that an RNA phage is likely to become the first organism about which nothing remains to be learned. It has been suggested that the replicase molecule is composed of two identical subunits, which would reduce the coding requirement, or that it is in part derived from host-coded protein. The surprising thing is not that the equation is as yet out of balance but that the present state of knowledge allows a balance sheet to be drawn up at all.

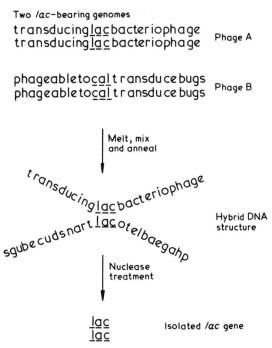

Figure 7.12 Scheme for the Isolation of Pure _Lac_ Operon DNA
For explanation see text.

7.3.3 Isolation of a 'gene'

The first isolation of a length of DNA corresponding to a
specific biochemical function (the _lac_ operon) was accomplished using two transducing phages [14]. Both, as a result of
genetic manipulation had acquired the _lac_ operon from their
bacterial hosts but inserted into their chromosomes in opposite
directions. Their DNA's were extracted and heated to separate
the strands of the double helices. These were then mixed and
annealed together, forming a hybrid structure as shown schema-
tically in Figure 7.12. Since the only region of homology was
the _lac_ region, this was the only double-stranded part of the
hybrid. The four single-stranded ends were then digested away
with an enzyme (fungal nuclease) specific for single-stranded
DNA, leaving the _lac_ operon in isolation. The process could

be monitored by electron microscopy which showed the four-ended hybrid clearly. The brevity of this account must not be allowed to impart a false impression of simplicity. This was a technological triumph of great complexity requiring skills and resources of a high order.

7.3.4 Grams of lac repressor

E. coli does not contain enzymes for fermenting lactose unless they are induced by the presence of that sugar because the *lac* genes are repressed by a protein (the *lac* repressor). Unfortunately this interesting material is present in normal cells in only vanishingly small amounts so that isolation of enough to study is very laborious. The isolation has been greatly facilitated by infecting the cells with a *lambda* phage that has lost its late genes (those responsible for lysing the host) in exchange for the *lac* genes of a previous host's chromosome and is, in addition, temperature-inducible. On induction by temperature-shift, many copies of the phage genome (including the *lac* gene) are made. The cell does not lyse so these serve as additional templates for the synthesis of *lac* repressor, resulting in such an increase in the amount formed that gram quantities are not nowadays remarkable.

7.3.5 Precellular evolution

Purified replicase from Qβ-infected *E. coli* can use Qβ-RNA as a template to synthesise copies *in vitro* when provided with a mixture of nucleotides and other defined chemicals. There is no doubt that the entire Qβ genome is replicated because when this new RNA is applied to *E. coli* spheroplasts (cells from which the wall has been removed), they become infected and complete, plaquable Qβ particles are formed. This replicase (or 'RNA- dependent RNA polymerase', to give it its full title) is specific for Qβ-RNA, as predicted by S. Spiegelman on the grounds that if it were not it would replicate all the multifarious RNAs found in the bacterial cell which would be of little value to the virus. Thus we have replication of macromolecules *in vitro,* a situation which, according to Spiegelman [15], 'mimics an early, precellular event, when the environmental selection operated directly on the genetic material rather than on the gene product.' One of the most mysterious phases in evolution thus became accessible to experimental investigation.

Spiegelman designed an experiment to see whether selective pressures in the test tube would evoke mutant nucleic acid molecules under conditions in which translation into protein did not occur. Essentially, the experiment consisted of serially diluting the reaction mixture in selective, unprimed (RNA-less) medium with such a short time interval between dilutions that the original RNA species was diluted to extinction, leaving only the faster-replicating mutants that arose. In one experiment the medium was made selective by the use of suboptimal nucleotide concentrations; in another an anti-metabolite drug was added. In both experiments mutant RNA's did indeed emerge, having replication rates substantially higher than that of the original when tested separately. Spiegelman suggests that the differences in replication rate may reflect different affinities for the replicase due to the different secondary structures assumed by the mutant RNA molecules. He further points out that there is a possibility of using similarly modified viral nucleic acids therapeutically, to compete with and arrest the replication of an established pathogen.

7.3.6 The bacteriophage model

The ease with which the bacterial hosts can be manipulated has allowed phage knowledge to get so far ahead of the rest of virology that it is natural for plant and animal virologists faced with a particular problem to look to related work with phages as a source of ideas. It is best to think of the bacteriophage model not in the sense of an ideal virus but as a basis for speculation. It is quite legitimate to speculate on the extent to which phages and other viruses *may* share a common property provided that such speculation, without evidence, is not mistaken for discovery. One may note, for example, that bacteria can acquire new properties on transduction or lysogenisation with phage and that animal cells can become cancerous on treatment with other viruses and ponder on whether a similar mechanism underlies both phenomena. At the moment, such evidence as exists suggests that the two processes are dissimilar — one cannot, for example, induce provirus in cancer cells as one can in lysogenic bacteria. The general term 'provirus' was coined many years ago but the only proviruses substantiated to date are prophages. Where the bacteriophage model is of considerable value is in the testing of new techniques and the teaching of existing methods if all that is required is a particle of viral composition and dimensions which is readily recognised and enumerated using the plaque assay. Centrifuges, filters and electron microscopes can all be proved with a phage preparation and for the biology teacher, phage may be the only available material for practical virology — the greenhouse, the Home Office licence, the CO_2 incubator, the candling box, all the cumbersome trappings of eukaryote virology that make the simplest demonstration time consuming and expensive, can be dispensed with and the student can learn the elements of virology with only test tubes, petri dishes and some simple media.

8 Applied aspects

8.1 Phage nuisances

Whereas phages are viruses of minor importance compared to those that infect Man and his domestic plants and animals, there are certain industries where bacteriophage infection is a nuisance, if not economically disastrous. These are industries that depend at some stage in their processes on changes brought about by specific bacteria.

8.1.1 Cheesemaking

In cheesemaking a culture of lactic streptococci known as 'starter' is first of all added to the milk. These bacteria convert the milk sugar (lactose) into lactic acid. The acidity of the milk is monitored closely during the process because the cheesemaker knows that unless the acidity has risen to the critical level for each subsequent stage of the process the cheese will be abnormal, possibly not of marketable quality if, indeed, it forms at all. Should bacteriophage able to attack the starter culture get into the vat the bacteria may be destroyed in a few hours and the acid level remains stationary. The cheesemaker refers to such an occurrence as a 'pack-up' because that is what he has to do. There is no point in adding more starter because, in destroying the first lot, the phage will have multiplied many thousandfold and the milk then

Figure 8.1 A Cheesmaking Vat
The curd is being cut. In cases of phage infection this stage may not be
reached. (Courtesy of Unigate Ltd.)

teems with phage ready to destroy subsequent additions even more quickly. The milk is thus wasted.

Two main remedies for the phage problem have been developed. One is to use, not a pure culture of bacteria but a mixture of several acid-producing strains. Since it is unlikely that any phage would have a sufficiently wide host range to infect them all, the usual result of phage infection is that only part of the bacterial population is eliminated and the process can proceed to completion, albeit slowly. It is just possible that the vat could become infected with several phages active against all the bacterial strains but such occurrences, fortunately, are rare.

The other solution is more sophisticated. Phage infection may occur through airborne contamination or infected apparatus associated with the cheese-making vat; it is much more serious if it occurs in the smaller vessel in which the bulk starter is grown. Most of the phages likely to be troublesome in cheesemaking require calcium ions for infection. This makes it possible to design a phage resistant medium (PRM) so that calcium is absent or held by chelating agents in an unavailable form.

Mixed starters and PRM may be employed together but there is still a need for good plant hygiene; careful disposal of whey, which may contain much phage, is important. The build-up of phage infection in a creamery can be a slow and insidious affair so that it is not noticed until it is firmly established.

8.1.2 Antibiotic fermentations

Certain antibiotics, of which streptomycin is a prime example, are produced by fermentation using actinomycetes. These are small filamentous organisms which have the same diameter as true bacteria and the same prokaryotic internal organisation. They are attacked by viruses which are often indistinguishable regarding their morphology and general behaviour from bacteriophages and are known as 'actinophages'. Actinomycetes require highly aerobic conditions and the actinophage used to gain access to the fermentation via the air supply.

Figure 8.2 Starter Culture Inoculation into a Bulk Starter Vessel
A shrouded, double-ended hypodermic needle is inserted through
rubber seals on both containers for aseptic transfer (Lewis's method).
(Courtesy of Unigate Ltd.)

It would then multiply rapidly, lysing or inhibiting the culture and causing serious production losses. Being very small, the actinophage could pass through the slag wool or carbon granule filters that were perfectly adequate for removing bacteria. The problem has been largely eliminated nowadays by improvement of the filtration media or use of heat sterilisation instead.

8.1.3 Solvent production

During the first World War and for some years afterwards, acetone and butanol, two important industrial solvents, were produced by a process using a bacterium, *Clostridium acetobutylicum,* to ferment sugary materials under anaerobic conditions. Bacteriophage was a frequent cause of slow fermentation and low yields. Phage was poorly understood in those days and one plant became so thoroughly infected with phage that production was no longer economic. It was abandoned and another plant erected many hundreds of miles away. The process is hardly used at all nowadays, most acetone and butanol being made chemosynthetically, but it is worthy of note as one of the first applied phage problems.

8.1.4 Baking and brewing

Several accounts have been published of a virus that infects yeast and may be the cause of the 'degeneration' of cultures used in these industries. It is called 'zymophage'. The evidence for it is tenuous.

8.2 Detection and identification of pathogenic bacteria

Phages have their positive side: although no longer contemplated as therapeutic agents, phages have an established place in medicine as diagnostic tools. Phage-typing is routinely used for the identification of pathogenic bacteria of which staphylococci and typhoid bacilli are prime examples. The techniques are delicate, requiring a high level of skill and judgement but in capable hands can distinguish between strains of bacteria inseparable by other methods. Analogies have been drawn between this and the fingerprinting of human beings.

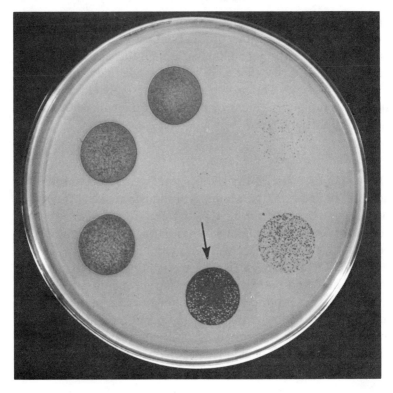

Figure 8.3 Determination of Routine Test Dilution (RTD)
Drops of phage dilutions increasing anticlockwise from 12 o'clock
were applied to the bacterial lawn. RTD (arrowed) is the last to show
confluent lysis. (Courtesy of Dr. E.A. Ashesov and Central Public
Health Laboratory, Colindale)

8.2.1 Salmonella typhi

Virulent strains of this organism may be identified using one
remarkable phage, the 'Vi-phage', which attacks many strains
but shows enhanced virulence towards the strain on which it
was last propagated. This phenomenon is known as 'host-
induced phage modification'. It is due to a phenotypic, not
genotypic change in the phage brought about by the methy-
lation of DNA bases in patterns that protect the phage DNA
from host enzymes that would otherwise destroy it ('restrictive

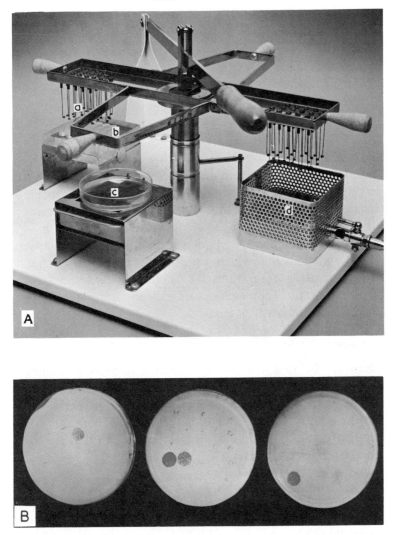

Figure 8.4 Mechanised Phage Typing
A. Lidwell's Applicator. The wire spirals a pick up phage suspension
from the wells in plastic block b and deposit it on the lawn of bacteria
to be typed in dish c. They are re-sterilized over the multiple gas-
burner d. B. Resulting patterns of virulence on three different bacterial
strains. (Courtesy of Dr. E.A. Ashesov and Central Public Health
Laboratory, Colindale)

enzymes'). The molecular basis is as yet far from clear but this does not detract from the practical value of Vi-phage. Stocks of Vi-phage, propagated on known strains, are maintained and, when applied to an unknown strain at the appropriate dilution, its identification is possible by matching patterns of virulence.

8.2.2 Staphylococcus aureus

No system comparable to Vi-phage is available for this pathogen; instead a large number of different phages, differing in their host ranges is used. They rarely attack only a single strain of staphylococcus so it is necessary to match patterns of virulence on a lawn of the unknown strain as shown in Figure 8.4. Each phage is applied at its Routine Test Dilution (RTD), which is that which just produces confluent lysis of the propagating strain. Lidwell's Applicator is a device for applying a set of typing phages quickly and accurately to many different lawns of bacteria.

8.2.3 Seed-borne plant pathogens

Phage can be used to detect the presence of bacterial plant pathogens in seeds. The seed is incubated in nutrient broth with a small, accurately known amount of phage specific for the pathogen being sought. If the phage is found subsequently to have multiplied, it may be inferred that the host pathogen was present.

8.3 Radiation dosimetry

An entirely different application for phage is as a measure of radiation dosage. The radiation sensitivities of certain phages, T2 for instance, are known very accurately. By including a known amount of phage in a material that is to be irradiated, or irradiating a phage control under identical conditions, the radiation dosage may be calculated from the residual infectivity. This is very valuable, giving a direct measure of the biological effectiveness of a radiation treatment that may defy calculation from physical and chemical data in complex systems.

9 Conclusion

The reader should by now have a fair idea of the nature and extent of bacteriophage biology; the way in which it is conducted; some of its achievements and ambitions. A selective bibliography is appended for those who wish to study it further. For anyone who has not yet read biology or biochemistry up to degree standard, Stent's 'Molecular Genetics' is strongly recommended, and Mathews' 'Bacteriophage Biochemistry' for the more erudite. After that, the phage literature being so vast, some measure of specialisation is advisable, if not already dictated by individual circumstances. The student may be fortunate enough to find a review of his particular field of interest in one of the specialised review journals such as 'Bacteriological Reviews', 'Annual Review of Microbiology' or 'Advances in Virus Research'. To keep abreast of current events it is advisable to read at least one virological journal thoroughly, 'Journal of General Virology' say or 'Virology' and to maintain surveillance of the rest by means of 'Current Contents', a journal that simply lists the titles of papers in other journals. Promising titles may be checked efficiently some months later in 'Biological Abstracts' or 'Virological Abstracts' to see whether the paper is worth reading in full. Finally one can find out precisely what is going on in a particular line by selecting a few recent papers of central importance and looking up in the 'Citation Index' who has subsequently cited them.

It is important for biology students and teachers, even those with no research ambitions, to maintain physical contact with the material of their philosophy. Priests of old claimed to derive superhuman knowledge from the inspection of entrails — Science now claims to have discovered the secrets of Life through the study of the invisible parasites of *E. coli,* the invisible inhabitant of Man's bowels! Unless the general run of students find that they can verify at least some of the doctrines of the phage-priests, Biology may become an edifice of sacred dogma based, like an ancient religion, on something that can be neither seen nor questioned except by the chosen few. The humblest plaque or dilution assay made by one's own hand is a confirmation of the credibility of our latter-day catechism; the present author hopes that he has shown the way to begin.

Appendices

APPENDIX A

UNITS OF MEASUREMENT IN ELECTRON MICROSCOPY

Linear dimensions of bacteriophages and comparable objects are given in nanometers (nm). This unit is superceding the Angstrom Unit (0.1 nm) and the millimicron (1.0 nm) for such purposes. 1 nm = 10^{-9} metres.

APPENDIX B

MEDIA FOR BACTERIOPHAGE STUDIES

Many special media have been described which give superior results in particular applications. Details of these must be sought in the relevant literature; here only general purpose media will be considered. A medium in which the host grows well is usually suitable for the propagation and assay of its phages provided that there is sufficient available calcium and magnesium and that the pH is not extreme. For coliphages the following medium is recommended:

	grams per litre
Nutrient broth powder (Difco)	8
Sodium chloride	5
Agar (Difco) for solid medium	15
(for sloppy medium	7)
Autoclaving: $121°$C for 15 minutes	

The above medium is unsuitable for organisms that must have carbohydrate, for example some streptococci. For these the following may be useful. It contains glucose and a buffer to prevent it becoming acid which does not render calcium unavailable.

Tryptone (Oxoid)	10
Yeast extract (Difco)	3
Glucose (bacteriological)	2
Sodium glycerophosphate 5½ H_2O (B.D.H. Laboratory Grade)	10
Tris-(hydroxymethyl)-aminomethane	1
Calcium chloride	0.25
Agar (Oxoid No. 3) for solid medium	10
(for sloppy medium	4)

The pH is brought to 7·8 with a few drops of dilute lactic acid. In a domestic pressure cooker 15 pounds per square inch (121°C) for 15 minutes is required for sterilisation. In larger autoclaves a shorter time, which must be determined experimentally, will be necessary to avoid charring the glucose; alternatively the glucose may be sterilised separately and added to the medium aseptically when cold.

APPENDIX C

THE PRINCIPAL PHAGE COLLECTIONS IN BRITAIN

The undermentioned organizations maintain stocks of phages for distribution to *bona fide* workers. A fee may be payable.

Phages of industrial, research and general interest
> National Collection of Industrial Bacteria,
> Torrey Research Station, P.O. Box 31,
> 135 Abbey Road,
> Aberdeen AB9 8DG,
> Scotland.

Phages of bacteria of medical importance
> National Collection of Type Cultures,
> Central Public Health Laboratory,
> Colindale Avenue,
> London NW9 5HT.

Phages of bacteria from milk and dairy products
    National Collection of Dairy Organisms,
    National Institute for Research in Dairying,
    Shinfield, Reading,
    Berkshire.

Phages of bacteria pathogenic to plants
    National Collection of Plant Pathogenic Bacteria,
    Plant Pathology Laboratory,
    Ministry of Agriculture, Fisheries and Food,
    Hatching Green,
    Harpenden, Herts.

APPENDIX D

SELECTED BIBLIOGRAPHY

1. Books on Bacteriophage and Related Subjects (an indication of
their nature is given where this is not evident from the title)
Adams, M.H. (1959) *Bacteriophages*, Interscience Publishers, Inc.,
    New York. (Deals with all aspects of the subject, somewhat out
    of date but still a valuable work of reference)
Cairns, J., Stent, G.S. and Watson, J.D. (Eds.) (1966) *Phage and the
    Origins of Molecular Biology*, Cold Spring Harbour Laboratory of
    Quantitative Biology, New York. (Gives an insight into the personal-
    ities and attitudes of some prominent phage biologists)
Cohen, S.S. (1968) *Virus-induced Enzymes*, Columbia University Press,
    New York and London. (The title is misleading since the book is
    almost entirely devoted to phage-induced enzymes).
Hayes, W. (1968) *The Genetics of Bacteria and their Viruses*, 2nd
    edition, J. Wiley and Sons, New York.
Hershey, A.D. (Ed.) (1971) *The Bacteriophage Lambda*, Cold Spring
    Harbour Laboratory, New York.
Mathews, C.K. (1971) *Bacteriophage Biochemistry*, Van Nostrand
    Reinhold Company, New York.
Reeves, P. (1972) *The Bacteriocins*, Chapman and Hall Ltd., London.
Stent, G.S. (1963) *Molecular Biology of Bacterial Viruses*, W.H. Freeman
    and Co., San Francisco and London. (A very readable and well
    illustrated account of the historical development of this subject
    from the very beginning)
Stent, G.S. (1971) *Molecular Genetics, an Introductory Narrative*,

W.H. Freeman and Co., San Francisco and London. (Phage occupies only a modest proportion of this book but is well seen against the background of genetics as a whole)

Tikhonenko, A.S. (1970) *Ultrastructure of Bacterial Viruses*, Plenum Press, New York, London.

2. Specific References, referred to by number thus [] in the text.

[1] Bachrach, U. and Friedman, A. (1971) Practical procedures for the purification of bacterial viruses. *Applied Microbiology* **22**(4), 706—715.

[2] Bradley, D.E. (1966) The fluorescent staining of bacteriophage nucleic acids. *Journal of General Microbiology* **44**(3), 383 — 391.

[3] Bradley, D.E. (1967) Ultrastructure of bacteriophages and bacteriocins. *Bacteriological Reviews* **31**, 230 — 314.

[4] Calendar, R. (1970) The regulation of phage development. *Annual Review of Microbiology* **24**, 241 — 296.

[5] Campbell, A.M. (1962) Episomes. *Advances in Genetics* **11**, 101 — 145.

[6] Kellenberger, E. (1972) Mechanisms of length determination in protein assemblies. *Polymerization in biological systems* (G. Wolstenholme and M. O'Connor, Editors) Ciba Foundation Symposium 7 (new series), 295 — 299.

[7] Fisher, W.D., Cline, G.B. and Anderson, N.G. (1964) Density gradient centrifugation in angle-head rotors. Analytical Bio-chemistry **9**, 477 — 482.

[8] Franklin, R.M., Hinnen, R., Schafer, R. and Tsukagoshi, N. (1974) Biochemical aspects of the structure and synthesis of bacteriophage PM2. *Proceedings of the Society for General Microbiology* **1**(2), 36 — 37.

[9] Kozak, M. and Nathans, D. (1972) Translation of the genome of a ribonucleic acid bacteriophage. *Bacteriological Reviews* **36**(1), 109 — 134.

[10] Luria, S.E. (1945) Genetics of bacterium — bacterial virus relationships. *Annals of the Missouri Botanic Garden,* **32**, 235 — 242.

[11] Luria, S.E. (1951) The frequency distribution of spontaneous bacteriophage mutants as evidence for the exponential rate of phage reproduction. *Cold Spring Harbour Symposium on Quantitative Biology* **16**, 463 — 470.

[12] Richards, K.E., Williams, R.C. and Calendar, R. (1973) Mode of DNA packing within bacteriophage heads. *Journal of Molecular Biology* **78**, 255 — 259.

[13] Rosenkranz, H.S. (1973) RNA in coliphage T5. *Nature* **242**, 327 − 329.

[14] Shapiro, J., MacHattie, L., Eron, L., Ihler, G., Ippens, K. and Beckwith, J. (1969) Isolation of pure *lac* operon DNA. *Nature* **224**, 768 − 774.

[15] Spiegelman, S. (1970) Extracellular evolution of replicating molecules. *The Neurosciences Second Study Program* (F.O. Schmitt, Editor) The Rockefeller Press, New York.

[16] Wyatt, H.V. (1974) How history has blended. *Nature* **249**, 803 − 805.

Index

Terms that occur widely in the text are defined or explained on the page given in heavy type.